Phuong Mai Dinh
Paul-Gerhard Reinhard
Eric Suraud

An Introduction to Cluster Science

Related Titles

Cundari, T.R., Boyd, D.B. (eds)

Reviews in Computational Chemistry, Volume 25
2 edn

2007
ISBN: 978-0-470-17998-7, Also available in digital formats

Barone, V. (ed.)

Computational Strategies for Spectroscopy
From Small Molecules to Nano Systems

2011
ISBN: 978-0-470-47017-6, Also available in digital formats

Komatsuzaki, T., Berry, R., Leitner, D.M. (eds)

Advancing Theory for Kinetics and Dynamics of Complex, Many-Dimensional Systems
Clusters and Proteins, Advances in Chemical Physics Volume 145

2011
ISBN: 978-0-470-64371-6, Also available in digital formats

Hadjichristidis, N., Hirao, A., Tezuka, Y., Du Prez, F. (eds)

Complex Macromolecular Architectures
Synthesis, Characterization, and Self-Assembly

2011
ISBN: 978-0-470-82513-6

Gonsalves, K., Halberstadt, C., Laurencin, C.T., Nair, L.S. (eds)

Biomedical Nanostructures

2008
ISBN: 978-0-471-92552-1

Vollath, D.

Nanomaterials
An Introduction to Synthesis, Properties and Applications

2008
ISBN: 978-3-527-31531-4

Hirscher, M. (ed.)

Handbook of Hydrogen Storage
New Materials for Future Energy Storage

2010
ISBN: 978-3-527-32273-2, Also available in digital formats

Grunenberg, J. (ed.)

Computational Spectroscopy
Methods, Experiments and Applications

2011
ISBN: 978-3-527-32649-5, Also available in digital formats

Rubahn, H.

Basics of Nanotechnology
3 edn

2008
ISBN: 978-3-527-40800-9

Phuong Mai Dinh
Paul-Gerhard Reinhard
Eric Suraud

An Introduction to Cluster Science

Verlag GmbH & Co. KGaA

Authors

Phuong Mai Dinh
Université Paul Sabatier
Laboratoire de Physique Théorique
Toulouse
France

Paul-Gerhard Reinhard
Universität Erlangen
Institut für Theoretische Physik
Erlangen
Germany

Eric Suraud
Université Paul Sabatier
Laboratoire de Physique Théorique
Toulouse
France

Cover Design
Physics meets art: The background of the cover represents a detail of a church window (© jorisvo/fotolia.com). The shining colors of church windows reflect their content in metallic particles, in modern words metallic nanoclusters.

Physics meets biology: illustrative example of gold nanoclusters attached to a DNA molecule. The presence of nobel metal nanoparticles in biological environments results in significant radiobiological and immunological effects the essential physical insights into which could be obtained through the molecular dynamics simulations, see http://www.mbnexplorer.com/.

All books published by **Wiley-VCH** are carefully produced. Nevertheless, authors, editors, and publisher do not warrant the information contained in these books, including this book, to be free of errors. Readers are advised to keep in mind that statements, data, illustrations, procedural details or other items may inadvertently be inaccurate.

Library of Congress Card No.:
applied for

British Library Cataloguing-in-Publication Data:
A catalogue record for this book is available from the British Library.

Bibliographic information published by the Deutsche Nationalbibliothek
The Deutsche Nationalbibliothek lists this publication in the Deutsche Nationalbibliografie; detailed bibliographic data are available on the Internet at http://dnb.d-nb.de.

© 2014 WILEY-VCH Verlag GmbH & Co. KGaA, Boschstr. 12, 69469 Weinheim, Germany

All rights reserved (including those of translation into other languages). No part of this book may be reproduced in any form – by photoprinting, microfilm, or any other means – nor transmitted or translated into a machine language without written permission from the publishers. Registered names, trademarks, etc. used in this book, even when not specifically marked as such, are not to be considered unprotected by law.

Hardcover ISBN 978-3-527-41246-4
Softcover ISBN 978-3-527-41118-4
ePub ISBN 978-3-527-67570-8
Mobi ISBN 978-3-527-67569-2

Cover Design   Simone Benjamin, McLeese Lake, Canada
Typesetting   le-tex publishing services GmbH, Leipzig
Printing and Binding   Markono Print Media Pte Ltd, Singapore

Printed in Singapore

Printed on acid-free paper

To Philippe, Elisabeth, and Evelyne

Contents

Preface *XI*

Units *XV*

1   **Clusters in Nature** *1*
1.1   Atoms, Molecules and Bulk *3*
1.1.1   Scales of Matter Down to Atoms *4*
1.1.2   More on Time Scales *5*
1.1.3   Binding in Atoms, Molecules and Bulk *7*
1.2   A New State of Matter? *12*
1.2.1   From Atom to Bulk, Small and Large Clusters *12*
1.2.2   Cluster Types *15*
1.2.3   Cluster Science *18*

2   **Measuring Clusters** *21*
2.1   Cluster Production *22*
2.1.1   Cluster Sources *23*
2.1.2   Sizes and Temperatures *27*
2.2   Excitations of a Cluster *30*
2.2.1   Collisions with Projectiles *31*
2.2.2   Laser Fields *32*
2.2.3   Coupling to Light and Optical Response *34*
2.3   Measuring Cluster Properties *36*
2.3.1   Mass Distributions *37*
2.3.2   Magnetic Moments *38*
2.3.3   Photon Signals *40*
2.3.4   Electron Signals *43*

3   **How to Describe Clusters** *47*
3.1   Approximations for the Ions *48*
3.1.1   The Adiabatic, or Born–Oppenheimer, Approximation *48*
3.1.2   Born–Oppenheimer Dynamics *49*
3.1.3   Beyond the Born–Oppenheimer Approximation *50*
3.1.4   Structure Optimization *51*
3.1.5   Approaches Eliminating Electrons *52*

3.2	Approximation Chain for Electrons	53
3.2.1	Overview of Approaches for the Electronic Subsystem	54
3.2.2	Ab initio Methods	55
3.2.3	Phenomenological Electronic Shell Models	58
3.2.4	Density Functional Theory	59
3.2.5	Semiclassical Approaches	64
3.3	Approximation Chain for the Ion–Electron Coupling	64
3.3.1	Pseudopotentials	65
3.3.2	Jellium Approach to the Ionic Background	67
3.4	Observables	70
3.4.1	Energies	70
3.4.2	Shapes	70
3.4.3	Stationary Response: Polarizability and Conductivity	71
3.4.4	Linear Response: Optical Absorption Spectra	72
3.4.5	Electron Emission	74

4 Some Properties of Free Clusters 77

4.1	Ionic and Electronic Structure	77
4.1.1	Magic Numbers	77
4.1.2	Shell Structure and Deformation	80
4.2	Optical Response	82
4.2.1	Basic Features	82
4.2.2	Impact of Deformation	84
4.2.3	Thermal Shape Fluctuations	86
4.2.4	The Width of the Mie Plasmon Resonance	87
4.3	Photoinduced Electron Emission	88
4.3.1	Total Ionization	88
4.3.2	Photoelectron Spectra (PES)	89
4.3.3	Photoelectron Angular Distributions (PAD)	91
4.4	Cluster Nonlinear Dynamics	93
4.4.1	Tunability of Lasers	94
4.4.2	On Ionization Mechanisms	95
4.4.3	Production of Energetic Ions and High Charge States	97
4.4.4	Variation of Pulse Profile	99

5 Clusters in Contact with Other Materials 101

5.1	Embedded and/or Deposited Clusters	101
5.1.1	On the Relevance of Embedded or Deposited Clusters	101
5.1.2	The Impact of Contact with Another Material	102
5.1.3	From Observation to Manipulation	107
5.2	On the Description of Embedded/Deposited Clusters	111
5.2.1	Brief Review of Methods	111
5.2.2	An Example of QM/MM for Modeling of Deposited/Embedded Clusters	114
5.2.3	A Few Typical Results	117
5.3	Clusters and Nanosystems	119

5.3.1	Towards More Miniaturization	*120*
5.3.2	On Catalysis	*121*
5.3.3	Metal Clusters as Optical Tools	*124*
5.3.4	Composite Clusters and Nanomaterials	*125*
6	**Links to Other Areas of Science**	*127*
6.1	Clusters in the Family of Finite Fermion Systems	*127*
6.2	Clusters in Astrophysics	*133*
6.3	Clusters in Climate	*137*
6.3.1	Impact of Clusters in Climate Science	*137*
6.3.2	From Aerosols to Water Droplets	*138*
6.3.3	Formation Mechanisms of Aerosols	*139*
6.3.4	Clusters as Seeds for Cloud Condensation Nuclei	*140*
6.4	Clusters in Biological Systems	*143*
6.4.1	Tailoring Clusters	*144*
6.4.2	Clusters for Medical Imaging	*146*
6.4.3	Clusters for Therapies	*150*
6.4.4	Nanotoxicity	*156*
7	**Conclusion**	*159*
	Further Reading	*161*
	References	*163*
	Index	*169*

Preface

Cluster science developed as an independent branch of science only a few decades ago. Since then its remarkable achievements have turned it into a major branch of science bridging the gap between microscopic and macroscopic worlds. Cluster science joins efforts from physicists and chemists and has led to impressive technological developments, opening the door to the nanoworld. It is now an established field of research with an impressive network of connections to neighboring scientific domains such as material science, but also to more remote ones such as, in particular, biology. We aim in this short book at addressing both these aspects, first, presenting cluster science as such, and second, indicating the connections to other fields of research.

Cluster science: a young field with a long history and a promising future Clusters were recognized in technological applications well before they were identified as physical objects. It was well known to Roman craftsmen and into the Middle Ages that immersing small pieces of a noble metal into a glass allowed for beautiful colors to then reside within the glass. More recently, photography became possible when realizing that small aggregates, namely "small" pieces of matter, of AgBr had a remarkable sensitivity to light, which with a proper chemical treatment, allowed to record and print images. However, over these many centuries, the idea of clusters as a subject of scientific research remained absent. One major reason was that one did not know how to isolate and to identify such microscopically small objects from a carrier environment, for example a glass in the Middle Ages, or more recently a gel. It has only been during the last quarter of the twentieth century that researchers have succeeded in producing in a controlled manner aggregates of various (controlled) sizes, thus opening the door to dedicated studies of clusters, also known as nanoparticles. For example, studies of clusters of various sizes allowed for the first time to systematically track the transition between atom/molecule and a bulk material. Within a few decades, cluster physics has become a lively domain of research, and has "invaded" many other domains where it is now fully admitted that clusters do play a major role. As typical examples, let us cite the many potential applications of nanosystems in material science in the race for miniaturization, as well as in medicine for drug delivery and imaging. We could also mention astrophysics with the composition of the interstellar medium, or climate science with aerosols. Cluster science has thus a promising future which motivates us to present its many facets in this book. Thereby we try to stay at an introductory level to address a broad readership.

Cluster science: a merger between physics and chemistry Clusters are constituted from atoms and interpolate between small molecules and bulk materials. As such, they thus call for expertise from chemistry – chemistry of small molecules but also solid-state chemistry – as well as from physics – atomic and molecular physics and solid-state physics. Indeed, the cluster community formed as a merger between various fields of physics and chemistry, including researchers involved in the study of other finite systems, such as nuclei or helium droplets. From this somewhat heterogeneous background emerged an original and rich scientific field primarily dedicated to the study of clusters themselves. Cluster science indeed developed over the last few decades into a somewhat specific domain. The study of clusters themselves made tremendous progress, reaching now in some cases a remarkable degree of detail, for example, even a time-resolved account of the dynamical response to a dedicated excitation. This high degree of detail was made possible partly because of the growing versatility of lasers over the same period of time and to the fact that clusters may have a strong coupling to light, especially clusters made out of metallic material. We will see many examples along that line throughout this book, both in the study of clusters themselves as well as in applications to other fields.

Cluster science: an interface between many domains A fascinating aspect of cluster science is that clusters play a role in several somewhat unexpected scenarios. This holds, for example, in astrophysics where it was recently realized that the composition of cosmic dust is "full" of clusters, whose influence on light signals received on earth may be crucial. But this also holds in terms of applications, for example, in drug delivery on specific targets in the human body. But they may be essential building blocks of new materials as well. The range of "applications" is thus enormous, from the largest times and distances in the universe to nanosized devices and materials, with an excursion into mesoscopic constituents of living cells in the human body. These apparently remote domains of scientific knowledge happen to share common objects, namely clusters. It is thus certainly an important issue to understand the properties of these fascinating objects.

The aim of this book is to introduce the reader to these many aspects of cluster science. The domain is huge and cannot be covered in depth within the limited size of the present book which should be an introductory text. It nonetheless indicates the wide range of cluster physics. We thus have confined the presentation to the basic aspects of clusters, being well aware that some aspects and many details are missing. This book is not meant to be an exhaustive review but rather a survey to motivate the reader to go deeper into the material. We have thus tried to supply relevant citations, mostly to textbooks or review papers and, when found helpful, to the proper specific citation. A strong underlying idea was to precisely cover general characteristics, often on a schematic basis, as well as some actual recent scientific results in order to enlighten the ever-developing nature of the field. The book thus consists of two parts of about equal size. The first half of the book (Chapters 1 to 3) includes a general introduction and provides the basic notions and keywords in experiments and theory. These notions should suffice for further reading of papers in the field. The last three chapters (Chapters 4 to 6) gather a collection of illustrative results. These chapters cover both properties of clusters themselves and their applications in various domains of science from astrophysics to material science and biology.

A book is always the result of numerous interactions with many colleagues. It is obvious that our project would not have converged without these many interactions. We would thus like to acknowledge the help of all these colleagues and tell them how much they helped, long ago or more recently, both in terms of science and personal contacts. We would, in particular, like to mention here: E. Artacho, M. Bär, M. Belkacem, D. Berger, G.F. Bertsch, S. Bjornholm, C. Bordas, M. Brack, F. Calvayrac, B. and M. Farizon, F. Fehrer, T. Fennel, G. Gerber, E. Giglio, C. Guet, B. von Issendorf, H. Haberl, J.M. L'Hermite, P. Klüpfel, U. Kreibig, J. Kohanoff, C. Kohl, S. Kümmel, E. Krotscheck, P. Labastie, F. Lépine, F. Marquardt, K.-H. Meiwes-Broer, B. Montag, M. Moseler, J. Navarro, V. Nesterenko, A. Pohl, L. Sanche, L. Serra, R. Schmidt, A. Solov'yov, F. Spiegelman, F. Stienkemeier, J. Tiggesbäumker, C. Toepffer, C. Ullrich, R. Vuilleumier, Z.P. Wang, P. Wopperer, F.S. Zhang, and G. Zwicknagel. Finally we would like to mention that this book emerges from a long-standing collaboration between the authors. This would not have been possible without the help of funding from the French–German exchange program PROCOPE, the Institut Universitaire de France, and the Alexander-von-Humboldt Foundation. We are thankful to these institutions to have supported us in our common efforts.

Toulouse
Erlangen
Toulouse

Phuong Mai Dinh,
Paul-Gerhard Reinhard and
Eric Suraud

March 2013

Units

We list here a few basic physical constants and units (data taken from [1]). We use the Gaussian system of units for electromagnetic properties (dielectric constant $\epsilon_0 = 1/4\pi$).

Electron mass: $m_e = 0.0156 \text{ eV fs}^2 a_0^{-2} = 0.5 \text{ Ry}^{-1} a_0^{-2} \hbar^2$
$= 1 \text{ Ha}^{-1} a_0^{-2} \hbar^2$

Light velocity: $c = 5670 \, a_0 \text{ fs}^{-1} = 274.12 \text{ Ry } a_0 = 137.06 \, a_0 \text{ Ha}^{-1}$

Fine structure constant: $\alpha = \frac{e^2}{\hbar c} = 0.007297 = \frac{1}{137.03}$

Charge: $e^2 = 1 \text{ Ha } a_0 = 2 \text{ Ry } a_0 = 14.40 \text{ eV Å} = 27.2 \text{ eV } a_0$

Bohr energy: $E_B = \frac{e^4 m_e}{2\hbar^2} = \frac{\alpha^2 m_e c^2}{2} = 13.604 \text{ eV} = 1 \text{ Ry} = \frac{1}{2} \text{ Ha}$

Bohr radius: $a_0 = \frac{\hbar^2}{m_e c^2} = 0.5291 \text{ Å} = 0.05291 \text{ nm}$
$= 0.5291 \times 10^{-10} \text{ m}$

Bohr magneton: $\mu_B = \frac{\hbar e}{2 m_e} = 5.788 \text{ eV T}^{-1}$

Boltzmann constant: $k_B = 8.6174 \times 10^{-5} \text{ eV K}^{-1}$

Energy scales: $1 \text{ Ha} = 2 \text{ Ry} = 27.2 \text{ eV}$
$1 \, h \text{ GHz} = 4.136 \times 10^{-6} \text{ eV}; \, 1 \frac{hc}{\text{cm}} = 0.1240 \times 10^{-3} \text{ eV}$

Time scales: $1 \text{ fs} = 10^{-15} \text{ s} = 1.519 \frac{\hbar}{\text{eV}} = 20.66 \frac{\hbar}{\text{Ry}} = 41.32 \frac{\hbar}{\text{Ha}}$
$1 \frac{\hbar}{\text{Ha}} = 0.5 \frac{\hbar}{\text{Ry}} = 0.0242 \text{ fs}$

Laser intensity: $I = \frac{c}{8\pi} |E_0|^2; \, \frac{I}{\text{W cm}^{-2}} = 27.8 \left| \frac{E_0}{\text{V cm}^{-1}} \right|^2$

Scale factors: $\hbar c = 1.9731 \times 10^{-7} \text{ eV m} = 1973.1 \text{ eV Å} = 274.12 \text{ Ry } a_0$
$\frac{\hbar^2}{m_e} = 1 \text{ Ha } a_0^2 = 2 \text{ Ry } a_0^2 = 7.617 \text{ eV Å}^2$

In dynamics, one simultaneously treats energy, distance and time scales, so that one has to consider proper combinations of these three quantities. Some standard packages are:

- eV, a_0 and fs
- Ry, a_0, $\hbar$ Ry^{-1} (1 $\hbar$ Ry^{-1} = 0.0484 fs)
- Ha, a_0, $\hbar$ Ha^{-1} (1 $\hbar$ Ha^{-1} = 0.0242 fs, called the atomic unit).

1
Clusters in Nature

Clusters, also called nanoparticles, are special molecules. They are composed from the same building blocks, atoms or small molecules, stacked in any desired amount. This is similar to a bulk crystal. In fact, one may view clusters as small pieces of bulk material. It has only been within the past few decades that clusters have come into the focus of intense investigations. During these few decades, cluster science has developed into an extremely rich and promising field of research. As often in science, technological applications of clusters existed before they were identified and understood. One of the most famous and oldest examples of the application of clusters in technology is the coloring of glass by immersing small gold clusters into the glass itself. The process allowed for some tuning of colors depending on the inclusions' size. This technology dates back to Roman times, where there is evidence craftmen had perfectly mastered this versatile technique. In scientific terms, such a phenomenon just reflects the size dependence of the optical response (that is, the color) of gold clusters in a glass matrix (although that prosaic formulation certainly does not give sufficient credit to the marvelous impressions attained that way). Another example of early applications is found in traditional photography which started about two centuries ago. The emulsion of a photographic film contains a dense distribution of AgCl (later AgBr) clusters whose special optical properties allowed to store information from light impulses and to visualize it later by chemical reduction. Progress in sensitivity and resolution was tightly bound in properly handling the cluster properties, where for a long time photographers did not even know that they were dealing, in fact, with clusters.

Figure 1.1 exemplifies clusters with an ancient and with a modern view. Figure 1.1a shows a church window from the St. Etienne cathedral in Bourges (France), whose impressive colors (not visible here, but which can be appreciated from the book cover) were fabricated to a large extent by Au clusters embedded in glass. Figure 1.1b was recorded with the most modern achievements of scanning tunneling microscopes (STM). It shows in detail Ag nanoparticles sitting on a highly oriented pyrolytic graphite (HOPG) surface. From the given length scale, we read off for this case cluster sizes in the range of a few nanometers, which corresponds to system sizes of about 100–10 000 Au atoms. As we will see, the combination of these quickly developing methods of nanoanalysis with nanoparticles, called clusters, constitutes a powerful tool for fundamental and applied physics.

An Introduction to Cluster Science, First Edition. Phuong Mai Dinh, Paul-Gerhard Reinhard, and Eric Suraud.
© 2014 WILEY-VCH Verlag GmbH & Co. KGaA. Published 2014 by WILEY-VCH Verlag GmbH & Co. KGaA.

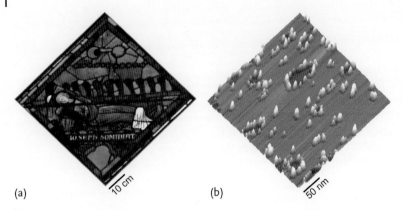

Figure 1.1 (a) Glass window of the St. Etienne cathedral in Bourges, France. Colors of church windows reflect their content in metallic particles. (b) Topography of silver nanoparticles deposited on (HOPG), recorded with an *in situ* scanning tunneling microscope (STM), from [2].

A first theoretical study which these days plays a basic role in cluster physics goes back to G. Mie in the early twentieth century [3]. Mie considered the question of the response of small metal particles to light, and how this optical response might depend on the size of the considered particle. It is interesting to quote Mie who turned out to develop a remarkable intuition of the future of cluster science: "Because gold atoms surely differ in their optical properties from small gold spheres", it would "probably be very interesting to study the absorption of solutions with the smallest submicroscopic particles; thus, in a way, one could investigate by optical means how gold particles are composed of atoms." This apparently simple and "reasonable" statement actually covers a large fraction of today's activities in cluster science as the interaction with light is a key tool for the investigation of cluster structure and dynamics.

In spite of Mie's intuition, the study of clusters as physical individual objects remained rather limited for the subsequent decades. Most investigations concerned clusters in contact with an environment (embedded or deposited). This limitation was due to the difficulty to create isolated clusters in a controlled manner. During the last quarter of the twentieth century, the capability of producing free clusters from dedicated sources finally triggered the emergence of cluster science on a systematic basis. The identification of the remarkable C_{60} clusters, the famous fullerenes [4], and the first systematic investigations of metal clusters [5] were impressive boosters. From then on, cluster science rapidly grew to an independent, although cross-disciplinary, field among the well-defined branches of physics and chemistry, ranging from fundamental research to applications in the context of nanotechnology.

Grossly speaking, clusters can be considered as large molecules or small pieces of bulk material. Their properties thus can be understood to some extent by methods from molecular and solid-state physics. Nonetheless, clusters represent a species of their own. One of the possibly most specific aspects of clusters is the fact that one can deliberately vary cluster size. Clusters are, so to say, "scalable" objects which bridge the gap between atoms/molecules and bulk material. This makes them useful testing grounds for the many-body problems, for example, to understand the path to bulk matter. They are true multidisciplinary objects staying in contact with many areas of science. This includes even such a remote field as astrophysics where clusters play a role in the formation mechanisms and the properties of cosmic dust. But the interest in clusters is not purely fundamental. We have already mentioned their early technological use in photography and artwork. Cluster research has meanwhile triggered a wide field of further applications. In chemistry, for example, scalability allows to play with the tunable surface to volume ratio which governs reactivity and may thus find key applications for catalysis. In material science, the discovery of fullerenes and nanotubes opened up new ways to design new materials [6] making carbon, and more recently "nano" science, an emerging field *per se*.

Cluster science with its many achievements now belongs to one of the most active fields in physics and chemistry, and offers, in particular through related fields, one of the fastest developing areas in applied as well as in fundamental science. One single and short book on cluster science can certainly not cover all aspects. We therefore confine ourselves to present key cluster properties and tools of investigation in a compact manner. The aim here is to give the reader a basic understanding and a motivation for further detailed readings. To reach this double goal (basics and motivation), we split our book in two major parts. The first half provides the essential concepts. The introductory Chapter 1 gives a first overview of the field. Chapter 2 introduces major experimental concepts, discussing cluster production and tools for the analysis of cluster properties. The subsequent Chapter 3 describes basic theoretical tools used in cluster science. These three first chapters, roughly representing half of the book, lay ground for all further reading. The second part tries to present selected examples superficially covering both, current cluster science and closely related fields. Chapter 4 presents in a compact manner some key properties of free clusters, both static and dynamic. It serves as a first application of the concepts introduced before. Chapter 5 concerns applications of clusters mostly in relation to material science. Finally Chapter 6 covers relations of cluster science to close domains such as astrophysics and biology. In the last two chapters, and to a lesser extent in Chapter 4, the topics to be covered are so vast that an extensive review is impossible in such a short book. We have thus chosen to select illustrative test cases which we discuss in some detail. We believe that such a strategy is more motivating and easier to grasp, though the selection may be somewhat subjective.

1.1
Atoms, Molecules and Bulk

Clusters have this remarkable and unique property to interpolate between individual atoms/molecules and bulk matter. Their size, in particular, can be deliberately chosen, so that this interpolation can be followed step by step if necessary. By being "in between" individuals and bulk, one could naively consider them as simple aggregates of individuals and/or finite pieces of bulk material. It turns out that they are neither the former nor the latter, and possess specific properties of their own. On the other side remain close relations with the extremes (atom/bulk). It is this subtle balance between specificity and connections with other systems that we first want to explore in this section. This provides a first overview of cluster science.

1.1.1
Scales of Matter Down to Atoms

We recall here a few basic scales of matter down to atoms and simple molecules, which in the following will constitute the building blocks of clusters. We focus on distances, energies and times. Figure 1.2 summarizes the following discussion in a compact and schematic manner indicating, in the time-distance plane, typical systems and observables of interest. The reader can also find a list at the beginning of this book summarizing the various units which are commonly used in atomic, molecular, cluster, and laser physics.

In bulk material, there is no clear macroscopic distance scale. We shall thus directly refer to the typical interconstituent distances, which lie in the nanometer and Å range. This also corresponds to the typical bond lengths in molecules. Atoms can be characterized by the typical radius of the electron cloud in the Å range. In the following, we shall thus use typical distances in the range 1 Å to 1 nm, and use as a standard unit Bohr's radius (hydrogen atom "radius") abbreviated as a_0 with $1\,a_0 \simeq 0.529$ Å.

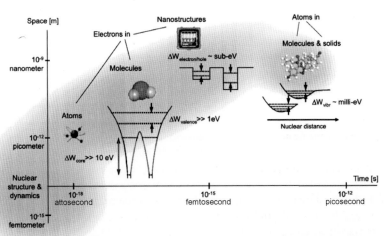

Figure 1.2 Schematic representation of typical time (abscissa) and length (ordinate) scales. We also indicate representative systems and components thereof, and associated energy scales for completeness. Today's investigations allow to access a wide range of distances and times in electronic systems including clusters, both from an experimental and a theoretical point of view. From [7].

Energies of interest span a much wider range with variations of six orders of magnitude. We shall take as a reference unit the electronvolt (1 eV = 1.6×10^{-19} J). It provides an appropriate unit for characterizing the binding of a valence electron (least bound electrons) to the system, often called the ionization potential (IP). The IP is typically a few eV varying, of course, from one system to the other, but within at most a factor of 10. The eV also roughly corresponds to the typical dissociation energy of a molecule or cohesive energy of a solid. Much lower energies, in the meV range, are associated to vibrational motion (atomic vibrations) in bulk and molecules. The meV is the typical energy range of rotational motion in molecules as well. This range of energies, though, is directly connected to atomic masses and nature of bonding. It thus shows larger variations reflecting the large span of atomic masses and electronic binding. On the side of higher energies, the binding of deeply bound electrons in atoms (and thus in molecules and bulk) ranges up to the keV domain. In most low-energy situations, such electrons remain safely bound to their parent nucleus. We shall therefore denote by "ion" the nucleus and the core electrons, to distinguish the latter ones from valence electrons. There however exist high-energy processes where core electrons can be excited, such as irradiation by high intensity laser fields or high-frequency photons (X-rays).

The above range of energies fixes a range of associated time scales on the basis of the energy-time uncertainty relation. The eV is closely related to the femtosecond (10^{-15} s) which appears as the natural time scale of valence electron motion in atoms, molecules and bulk. The meV range in turn is associated to the picosecond domain, characteristics of atomic motion. On the other extreme, keV electrons are typically associated to attoseconds (10^{-18} s). This very short time scale was not accessible until very recently. The advent of new light sources now allows to produce light pulses with durations down to a few tens of attoseconds. Such short pulses suffice to directly excite low lying, deeply bound electrons. In a way similar to energies, accessible time scales today thus span several orders of magnitude in the domain of interest of cluster science.

1.1.2
More on Time Scales

Figure 1.2 provides a very general overview of typical time scales of cluster properties, cluster dynamics, and excitation mechanisms. Clusters are built of atoms and thus naturally cover the various time scales associated both to individual atoms and assemblies thereof. Furthermore, as already mentioned in Figure 1.1, the possible contact with an environment (surface, matrix, solvent, ...) may introduce more time scales to the picture. Finally, it is interesting to detail a bit the typical times scales associated to lasers, which, as we shall see at several places throughout the book, constitute a basic tool of investigation of cluster properties. We thus present in Figure 1.3 a schematic but more detailed overview of times related to electronic and ionic motion, lifetimes for relaxation processes, and laser characteristics. Numbers are given in the case of metal clusters because specifying the cluster type allows to be more specific on time scales. Changing the nature of the cluster would

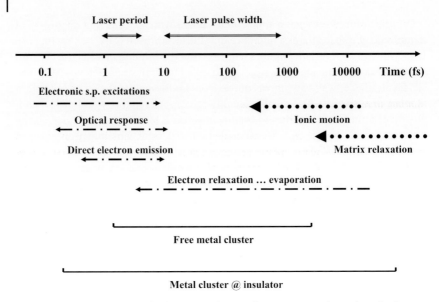

Figure 1.3 Typical time scales for the various dynamical processes in a cluster, here for the example of free sodium clusters, and clusters in contact with an inert, insulating substrate.

probably alter the detailed numbers, but not the qualitative relations between the various time scales. The fastest cluster time scales are related to electronic motion. Core electrons have cycle times of 0.1 fs and faster (depending on the element). The Mie plasmon period, see Eq. (2.6), most important for metal clusters, is on the order of femtosecond, see Section 2.2.3. In the same range, but with wider span from subfemtosecond to several femtosecond, are cycle times for other single-particle excitations and direct electron escape, that is single-particle excitation directly into the continuum. Somewhat slower is the plasmon decay due to Landau damping, a mechanism which is well known from plasma physics [8], see also the discussion of Figure 4.7. The most widely varying electronic times are related to damping from electron–electron collisions and thermal electron evaporation. Both strongly depend on the internal excitation energy of the cluster. They are long for low and moderate excitations, as indicated in Figure 1.3. Ionic motion spans a wide range of slow time scales. Vibrations, which may be measured by Raman scattering, typically in the meV regime, have cycle times of order 100 fs to 1 ps. Strong laser irradiation can lead to Coulomb explosion of the cluster where ionic motion becomes somewhat faster. Besides the ionization effects, the thermal relaxation between electrons and ions takes much longer, up to the nanosecond range. Ionic relaxation processes are even slower, for example thermal emission of a monomer can easily last μs. As indicated in Figure 1.3, the pulse duration of optical lasers may be varied over a wide range and extends in principle from fs (even some hundreds of attoseconds) to ps or even ns.

The above discussion reveals that cluster dynamics comprises a huge span of time scales which are extremely hard to treat at the same level of refinement. A most basic and widely used approximation is to treat nuclei, or ionic cores, respectively, as classical particles while the electrons remain quantum mechanical particles. We will come back to that aspect in more detail in Chapter 3.

1.1.3
Binding in Atoms, Molecules and Bulk

Atoms, molecules and solids are all made from the same constituents, namely nuclei and electrons. And yet, matter provides a great variety of properties. The reason is that electron binding shows up in very different manners, depending on the elements involved and the composition thereof. In this section, we will briefly discuss the different mechanisms of electron binding.

1.1.3.1 Atoms

Atoms consist, as is well known, of a nucleus at the center surrounded by a cloud of electrons. They interact with each other and with the central nucleus mainly through Coulomb interaction. The immense attraction provided by the positively charged nucleus suffices to counterweight the repulsion between electrons themselves. The basic theory is known as the central field approach [9]. Each electron independently moves in a common mean field, the central field, which consists of the Coulomb field of the central nucleus augmented by the screening field from the other electrons. The predominantly spherical shape of the central field leads to a pronounced bunching of the single-electron spectrum into shells, each one containing a couple of degenerated states. The Pauli principle determines the filling of these shells with increasing electron number. This simple shell model picture provides the key to a first sorting of atomic structure. Particularly important are shell closures which arise for those electron numbers where the occupied shells are all exactly filled. In the Mendeleev classification, this situation corresponds to the rare gas atoms He, Ne, Ar, Kr, Xe, and Rn. These are particularly inert in chemical reactions. In general, the amount of shell filling determines the chemical properties. Extreme cases are those next to shell closure. The alkalines Li, Na, K, Rb, Cs, and Fr, have one single electron on top of a closed shell and are considered as simple metals. At the other side of shell filling are the halogens F, Cl, Br, I, and At which have an almost complete shell, missing just one electron. Both these groups are particularly reactive. One can go stepwise further, two additional electrons or two missing electrons, and so forth. The properties of some transitional atoms may require to watch not only the shell of least bound electrons but also the next lower shell. This is, for example, the case for the transition metals Cu, Ag, and Au. All in all, the ensemble of these "active" electrons constitute what one usually calls the valence electrons, namely those which determine binding structure and low-energy dynamical properties.

The calculated shell structure of two typical metals is depicted in Figure 1.4. The states are labeled in standard atomic physics notation. The letters denote the orbital angular momentum l with the series s, p, d, f, g, ... corresponding to $l =$

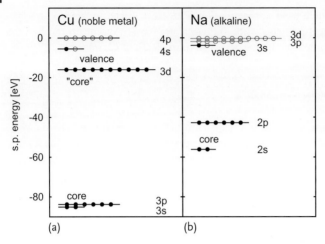

Figure 1.4 Comparison of the calculated level sequence in a noble metal atom, Cu (a), with a simple metal atom, Na (b). Single-particle (s.p.) energies are indicated by horizontal bars. The naming of the states (atomic convention) is given to the right near each level line. Occupied states are indicated by filled circles and unoccupied ones by open circles. Deeply lying states are not shown. All results have been computed by density functional theory (see Section 3.2.4).

0, 1, 2, 3, 4, ... The numbers count the occurrence of an angular momentum in increasing order of energies. For example, the 4p states denotes the fourth $l = 1$ state and 3s the third $l = 0$ states. Figure 1.4a shows the atomic level sequence for the noble metal Cu, and Figure 1.4b shows that for the simple metal Na. Note that there are several lower states (1s for Na and 1s, 2s, 2p for Cu) which are so deeply bound that they fall far below the bounds of the plot. For the remaining states, one still sees a clear distinction. The 3s, 3p states for Cu and the 2s, 2p states for Na are far away from the least bound states (which constitute the Fermi surface of the system). These are rather inert to bonding and low-frequency dynamics, and are thus considered as core states building together with the nucleus the ionic core. There remains one occupied valence state, the 3s for Na and the group 3d, 4s for Cu. The distinction between core and valence electrons is, of course, a matter of decision and may differ under different circumstances. For example, one may put the 3d state in Cu into the core states for low-energy situations where the 4s–3d gap of about 10 eV is sufficiently large. On the other hand, highly dynamical processes (involving high energies and/or frequencies) may require one to include the 3d state in Cu amongst the active electrons.

Quantitative indicators of the chemical properties of an atom are the binding energies of the lowest unoccupied molecular orbital (LUMO) and of least bound electrons known as highest occupied molecular orbitals (HOMO). The latter energy is directly measurable as the ionization potential (IP), that is the energy which is needed to singly ionize the atom. For example, rare gas atoms are associated with particularly large IP which indicate a high resistance against ionization (and chemical reactions in general). The binding energy of the LUMO is related to the electron

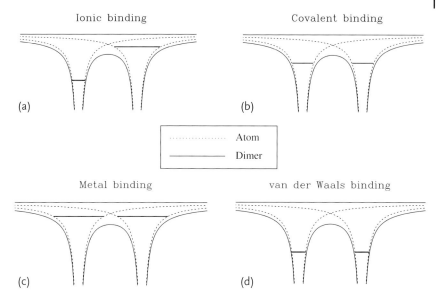

Figure 1.5 Schematic view of potentials and valence levels in dimers with the four basic types of binding as indicated (a–d). The potential seen by an electron in the molecule is drawn with solid lines, while those originating from single atoms are represented by dashes. The valence levels are indicated by horizontal straight lines.

affinity which determines whether an atom is an electron donor (e.g., alkalines) or acceptor (e.g., halogens).

1.1.3.2 Simple Molecules

Because chemical binding is by nature a low-energy phenomenon, involving at most a few eV energy, the properties of the valence electrons mainly determine how atoms bind together. The Mendeleev classification thus provides a gross map of which atom may possibly bind with another one. We first discuss the dimer molecule which is the simplest conceivable case. The question is how electrons will rearrange themselves between both atoms. The answer is to a large extent related to electronic energies. Strongly bound electrons will tend to remain on their parent atom, while loosely bound ones will have a tendency to leave it. Depending on the partner atom, this leads to different situations, as sketched in Figure 1.5.

In the case of ionic binding (Figure 1.5a), one loosely bound electron is transferred from its parent atom to a host atom offering an empty, more deeply bound level. The transfer is all the easier the smaller the IP of the parent atom and the higher the electron affinity of the host. The electron then predominantly localizes on the host atom: the donor atom becomes positively charged and the host one negatively, leading to a robust binding between the two such formed ions. Typical examples of ionic binding are realized between alkaline (electron donor) and halogen (electron acceptor) atoms such as, for example, in NaCl, NaI, or LiF.

Ionic binding is all the more likely when the energy difference between the valence shells of the two atoms is large. In other cases, electron transfer becomes unfavorable and electrons tend to remain closer to their parent atoms. Binding then stems from charge sharing rather than charge transfer: the valence electrons tend to form a common electronic cloud establishing the binding of the two atoms together. Such a picture traditionally leads to two kinds of binding, depending on the actual strength of binding of each electron on its parent atom. If the valence electrons are weakly bound and if the atoms become sufficiently close to each other, the electronic wave functions become delocalized and fill the space all around the two atoms leading to a so-called metallic bond (Figure 1.5c). Alkalines are the typical elements which establish metallic bonds between each other. Metallic binding is especially simple in alkalines as it involves only one loosely bound electron per atom, hence the notion of "simple" metals to characterize these systems. Metallic binding is also observed in more complicated metals such as, for example, Cu, Ag, Au, or Pt. The shell below the HOMO is taking part in the binding. In this case, it is the d shell energetically close to the valence s shell (see Figure 1.4a). The energetic neighborhood of the d shell adds strong polarization effects with impact on binding and dynamical properties.

When electrons are initially more strongly bound to their parent atom, electrons rather gather in the region of smallest potential energy, that is between the two partner atoms, instead of fully delocalizing themselves. This leads to the so-called covalent bond realized between atoms like C or Si (Figure 1.5b). However, it should be noted that covalent and metallic bonds as described above are idealized situations. There is a smooth transition between these extremes. In addition, it is usually hard to disentangle these two types in small molecules. The distinction becomes better defined in bulk material.

In the extreme case where valence electrons are too deeply bound in the atom, as in rare gas atoms, neither charge transfer nor charge sharing are possible between the two atoms. Electrons remain basically localized to their parent atom. Still, the electronic cloud of each atom is influenced by the partner atom by mutual polarization. This polarization of the electronic clouds has to be understood as a dynamical correlation effect, with constantly fluctuating dipoles, rather than as a mere static polarization. It results in mutual dipole–dipole interactions between the two atoms, which finally establishes a binding of the system. Although the resulting binding is much weaker than the previous ionic, metallic or covalent bonds, it suffices for the binding of rare gas molecules up to possibly large compounds. This type of binding is known as van der Waals (or molecular) binding, in reference to the interactions between two neutral atoms or molecules (Figure 1.5d). It primarily concerns binding between rare gas atoms but will also be useful to describe binding between small molecules such as CO_2, because well-bound molecules as such represent also closed shell situations.

Simple energetic considerations for HOMO and LUMO thus allow one to identify four major types of bonding between atoms, as schematically represented in Figure 1.5. These energetic considerations are directly linked to the degree of localization of the electrons binding the two atoms: the more bound the electrons, the more localized their wave functions. As already pointed out, there is no clear separation between the various binding mechanisms, but rather a continuous path between them. Still, the sorting in four classes provides useful guidelines for understanding the binding of most molecular systems. As a final remark, one should note that some other "classes" of binding are sometimes introduced such as the separation between van der Waals (for rare gas) and molecular (for molecules), and that between covalent (generic one) and hydrogen binding (specific to covalent binding involving hydrogen atoms). Such fine distinctions do not basically alter our global classification scheme and we shall thus ignore such details in the following.

1.1.3.3 Bulk

We have classified binding on the basis of binding energies of the valence electrons. Although bulk spectra differ by nature from the discrete atomic spectra, it turns out that binding proceeds in a similar way and finally allows a similar classification.

In a solid, the infinity of connected atoms smears discrete energy levels into bands of continuous energy levels. These bands are separated from each other by energy gaps. Band structure is related to the width of the gaps, that of the bands, and where they are located in the energy spectrum. One distinguishes between two extremes, conductors and insulators. In the former ones, the valence bands containing the least occupied electronic states overlaps with the conduction band of empty electronic states so that electrons can switch empty conductance levels at no energetic cost. Such a system is ideally represented by simple metals such as alkaline metals in which, again, electronic wave functions delocalize over the whole system. The other extreme are insulators in which valence and conduction bands are separated from each other by a finite, mostly large, energy gap. There is, of course, a continuous path between these two extremes depending on the size of the energy gap. The transitional cases (small gaps) are known as semiconductors where conductivity can be easily induced by thermal agitation or dedicated doping [10].

Similar as in molecules, bonding in non-metals, that is insulators, covers different forms, essentially reflecting the degree of localization of electrons around their parent atom. In covalent crystals, electrons are semilocalized, gathering along the lines joining atoms together. The typical example of such a binding is the case of diamond. The last two classes of insulators leave electrons fully localized on atomic sites. In van der Waals crystals such as a solid noble gas, electrons basically remain bound to the original atom they were attached to. Just as in simple molecules, they are bound by polarization effects. Finally, ionic crystals are composed of pronounced donor atoms (metals) and pronounced acceptors (for example halogens) in a regular manner. This leads to ionic bonds where the electrons are localized on atomic sites thus producing an interlaced lattice of cations and anions. Typical examples are alkaline-halogen compounds again, such as NaCl, NaF, or LiF.

It is not surprising that the classification of bonding in solid-state physics matches that of molecular physics. And again, one should emphasize that the four types of bonding are idealizations and that reality often resides in between. Clusters provide the unique opportunity to map a continuous path from atom to bulk as we will discuss in the next section. One of the interesting aspects consists in finding situations where the nature of bonding changes with system size.

1.2
A New State of Matter?

1.2.1
From Atom to Bulk, Small and Large Clusters

By construction, clusters are aggregates of atoms or molecules with arbitrarily scalable repetition of a basic building block and with an intermediate size between atoms and bulk. It corresponds to a chemical formula of the form X_n, where $3 \lesssim n \lesssim 10^{5-7}$ and X stands for an atom or a simple molecule as, for example, H_2O. The actual upper limit in size is given by experimental conditions. The size at which bulk-like behavior appears depends on the considered observable.

One is tempted to argue that clusters are "nothing else" than "big" molecules or "small" pieces of bulk. This is not the whole story. Indeed, clusters can develop properties of their own which differ from molecules and bulk. In order to illustrate that point, let us give a few typical examples. One of the characteristics of clusters is the usually huge number of isomers they possess, even in a small energy window. Already for such a small cluster as Ar_{13}, one can find hundreds of isomers. Metal clusters are also swamped by isomers, because of the softness of their binding. This behavior differs from the molecular case where the number of isomers is usually small, of order a few units at most. This exorbitant isomerism results from the homogeneous constitution of clusters. In other words, the ground-state potential energy surface (or Born–Oppenheimer surface, see Section 3.1.2) is in general very flat, which makes it very hard to figure out the actual ground-state structure of the system. It is therefore very hard to assess which is the most stable structure, in measurements as well as in numerical calculations. Standard quantum chemistry techniques are well adapted to molecules with few isomers, but are here often at a loss. The point is even more dramatic as clusters are usually formed at finite temperature (see Section 2.1.1). In this case, a given cluster covers a huge variety of isomers (thus shapes) in its equilibrium ensemble (see Section 4.2.3).

We now have seen that clusters are not simply large molecules. They are not finite pieces of bulk either. This is, of course, particularly true for small clusters (some tens of atoms) for which the "bulk" limit is obviously far away. But this remains also true for clusters containing several thousands of atoms. There are various features which distinguish a small piece of bulk from bulk itself. A basic difference lies in the electronic level structure which exhibits bands in bulk but discrete levels in finite systems. The path from a fully discrete molecular level scheme to continuous bands is a smooth process, as sketched in Figure 1.6. Figure 1.6b provides a schematic picture of the evolution of electron levels with cluster size, introducing some basic terminology from molecular and solid-state physics.

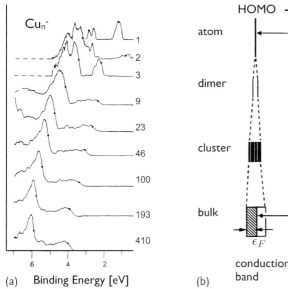
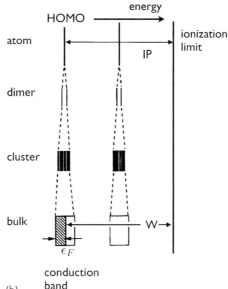

Figure 1.6 (a) Evolution of single-electron spectra with system size for a series of Cu_n^- cluster anions. The spectra are deduced from photoelectron spectroscopy (see Section 2.3.4). The upper atomic level in Cu is the 4s state and the lower one the 3d, see also Figure 1.4. One can track down how the corresponding bands develop out of the atomic states. The points in the spectra for larger clusters indicate the upper band edge. (b) Schematic view of the development of bulk bands out from atomic levels. After [11].

The ionization potential (IP) becomes the work function (W) in solids. Up to a sign, the IP is the energy of the HOMO. The energy difference between HOMO and LUMO (see Section 1.1.3.1) becomes the "energy gap" in solids. The scheme nicely illustrates how the density of electron states (DOS) quickly increases with increasing system size. The view is complemented by the experimental example in Figure 1.6a showing the spectral distribution of occupied states in a series of anionic Cu_n^- clusters. The measurements [11] were performed using photoelectron spectroscopy which will be discussed in Sections 2.3.4, 3.4.5.2, and 4.3. For small clusters, one can still resolve fine peaks representing single states. The patterns are quickly smoothed for heavier clusters. There are also small, but systematic energy shifts on the way to bulk. Note that energy bands in solids are not necessarily gathering around the atomic levels where they originate from.

A way to quantify finite size effect characteristics of clusters (hence at variance with molecules and bulk) is the surface to volume ratio. Most clusters are indeed large enough that one can distinguish a surface zone, and yet, they remain sufficiently small such that a sizable fraction of the constituents lie on the surface of the system, at variance with bulk. For example, the Ar_{55} cluster has 32 atoms on its surface which leads to a ratio of surface atoms to total is thus $32/55 \sim 0.6$. On the other side, a cubic piece of bulk material of 1 mm^3 volume contains about $(10^{-3}/10^{-10})^3 \sim 10^{21}$ atoms, out of which about $(10^{21})^{2/3} \sim 10^{14}$ lie at the bound-

Table 1.1 Schematic classification of clusters according to the number n of atoms: diameter d for Na clusters (second row), estimate of the ratio of surface to volume atoms f (surface fraction, in third row). After [12].

Observable	Very small clusters	Small clusters	Large clusters
Number of atoms n	$2 \leq n \leq 20$	$20 \leq n \leq 500$	$500 \leq n \leq 10^7$
Diameter d	$d \leq 1\,\text{nm}$	$1\,\text{nm} \leq d \leq 3\,\text{nm}$	$3\,\text{nm} \leq d \leq 100\,\text{nm}$
Surface fraction f	Undefined	$0.9 \gtrsim f \gtrsim 0.5$	$f \leq 0.5$

aries, hence a surface to volume ratio of 10^{-7}. Taking a sample in the micrometer range increases the ratio to about 10^{-6}. It is only for typical cluster sizes in the nanometer range that one recovers ratios of order unity. Size is thus obviously a key parameter in cluster science and a classification in terms of the fraction of surface atoms to volume as illustrated in Table 1.1. Of course, this classification is schematic, and there are no strict boundaries between the various classes of clusters. But the scheme helps to sort the sizes and to specify the terms "small" or "large" clusters used in the following.

Another way to visualize how clusters connect between atoms/molecules and bulk is to plot a given observable as a function of cluster size n. We shall see examples of such trends at several places in the book, see, for example, for magnetic moments in Figure 2.6, for IP in Figure 4.1, or for X-ray yield in Figure 2.8. Figure 1.7 shows the evolution of the cohesive energy per atom for alkaline clusters. The cohesive energy of a cluster with n atoms is composed of the sum of dissociation energies from the atom up to the given n [13]. This definition merges for very large n into the cohesive energy of the corresponding bulk system [10]. The trends towards bulk are best visualized when plotting versus $n^{-1/3}$, which is proportional to the inverse cluster radius. The point at $n^{-1/3} = 0$ shows the bulk value. Average trends of almost any observable $X(n)$ having a finite bulk value X_0 can be expanded as $X(n) \approx X_0 + X_1 n^{-1/3} + X_2 (n^{-1/3})^2 + \ldots$ where X_1 stems from planar surface effects, while X_2 is associated with curvature. Quantum shell effects may produce fluctuations about these smooth average trends. The straight lines in the insets indicate the lowest order trend for the cohesive energy $E_n \approx E_{\text{vol}} + E_{\text{surf}} n^{-1/3}$. Neutral clusters follow soon this line with only small fluctuations about it. Cluster cations approach this line only for larger system sizes. The example of Figure 1.7 demonstrates the trends towards bulk for the cohesive energy in alkalines. Actual convergence to bulk depends on the observable, on the material, and on the resolution of a measurement. For example, the experimental resolution of photoelectron spectra place the transition from well-separated discrete electron levels to quasi-continuous bands at about $n \approx 100$ in Cu_n^-, see Figure 1.6. Electronic shell effects, as for example HOMO-LUMO gaps at shell closures, proportionally shrink to $n^{-1/3}$. And yet, their importance has triggered large efforts to resolve them up to the range of $n \approx 3000$, while atomic shell effects have been resolved up to $n \approx 10\,000$ (see Section 4.1.1). The peak frequency of optical response also converges towards its

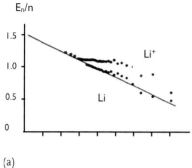
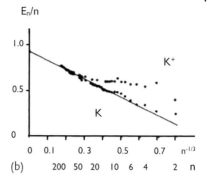
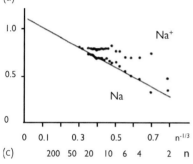

Figure 1.7 Cohesive energy per atom for neutral and cationic clusters of Li (a), K (b), and Na (c) drawn versus $n^{-1/3}$ where n is the number of ions in the cluster. The neutral clusters are approaching nicely a straight line from which one can read off the volume energy (intercept at $n^{-1/3} = 0$) and surface energy (slope). Adapted from [13].

bulk value with a linear term in $n^{-1/3}$. This means that colors keep drifting with size up to very large clusters in the range above 10 000. Tuning cluster size was, in fact, the mechanism to tune colors in Roman and medieval glass technology, see Figure 1.1.

Altogether, we see that clusters are distinguished from molecules and bulk by a couple of finite size effects: influence of surface and its curvature, atomic shells, and quantum effects from electronic shells. More examples will be found throughout the rest of the book.

1.2.2
Cluster Types

We have seen above that there are four major classes of binding: metallic, covalent, ionic, and van der Waals (or molecular). They exist much similar in molecules and bulk. Accordingly, we identify the same four types of binding in clusters. Figure 1.8 gathers typical examples of these four classes of clusters. The famous C_{60} fullerene (Figure 1.8d) provides a beautiful example of covalent bonding in clusters. The cluster exhibits here a well-defined atomic structure with electrons localized along the various links between the atoms. Covalent binding is typical in clusters made of atoms belonging to the carbon group (carbon and silicon in particular). The case of Ar_{13} (Figure 1.8c) provides an example of a van der Waals cluster, as it is found for all rare gas clusters. This particular example Ar_{13} corresponds to a closed atomic shell (see Section 4.1.1). The atomic structure of rare gas clusters can be computed rather simply by effective atom–atom potentials [10]. Here, we show the result of a quantum mechanical calculation to also illustrate the electronic structure. Each atom clearly carries its own electron cloud which has little overlap with the electrons from other sites. Binding is dominated by long-range van der Waals forces. An example of ionic binding is the Na_4F_4 cluster (Figure 1.8b). It is an especially simple case associating "ideal" partners (alkaline + halogen) in an almost stoichiometric manner. Four electrons move from Na to F. This yields four Na^+ cations without valence electrons and four F^- anions with rather well-localized electron charge. The binding is dominated here by the Coulomb interaction between positive and negative charges. It is worth noting that a regular structure already appears in such a small cluster. Finally, metallic bonding is illustrated on the example of Na_8 (Figure 1.8a). The electronic density extends more or less smoothly over the entire system. It is typical for metals that the valence electrons belong to the system as a whole and become independent from single ions. This makes them the ideal testing ground for electronic shell effects, see Section 4.1.1 for magic numbers, and Section 4.1.2 for deformation driven by the electron cloud. Note that the figure is drawn using the same scale for all four clusters. This gives an impression of the relative sizes. The binding in rare gas clusters is particularly weak, leading to soft and extended clusters. The opposite situation of very compact binding is found in ionic clusters as here the case of Na_4F_4. Covalent binding is still rather compact (note that the C_{60} embraces more atoms than the other examples in Figure 1.8). Metallic binding tends to be slightly softer.

A summary of the four classes of bonding and some consequences thereof is presented in Table 1.2. The least bound clusters are van der Waals (or molecular) clusters. They can be considered as a collection of weakly interacting atoms. On the contrary, the strongest binding is usually attained in ionic clusters. Covalent clusters may bind almost as strongly. Metal clusters are generally a softer bond than covalent ones, thus constituting an intermediate class between the almost unbound van der Waals clusters and the tightly bound ionic (or covalent) clusters. A word of caution is necessary here. Such sorting schemes rely on idealizations and reality often falls in between the categories. One can even find examples where the type of binding depends on the cluster size, as for example in Hg_n^-. Indeed photoelectron spectra measurements indicate that these clusters are rather covalent for $n \leq 13$, then that they exhibit semiconductor-like densities of states for intermediate sizes, and a closure of the band gap for $n \geq 300$ and thus a metallic behavior.

Before concluding this section, it is interesting to make a remark on scales. Whatever the bonding type, binding energies lie in the eV range and elementary bond lengths take values in the few a_0 range, which perfectly matches our initial energy and distance scale (see Section 1.1.1). Complementary to this is that it is interesting to also evaluate a typical force in a cluster. This allows to compare with external fields probing the system (laser or colliding projectile) and so, to estimate regimes

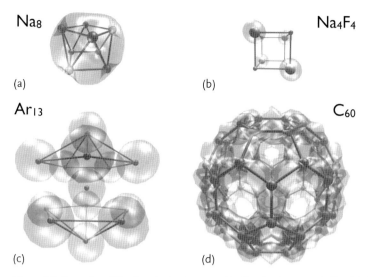

Figure 1.8 Examples of the four cluster types according to binding. Ions are indicated by small balls and the electron cloud by transparent gray. Nearest atoms are connected by bars to better visualize the atomic structure. All four clusters are plotted with the same scale, allowing a direct comparison of sizes. Metallic binding for the example Na_8 (a). Ionic binding for the example Na_4F_4, where Na atoms are denoted by small spheres and F atoms by larger spheres (b). Van der Waals binding for the example Ar_{13} cluster computed with effective atom–atom potentials (c). Covalent binding for the example C_{60} (d).

Table 1.2 Classification of the four types of binding in clusters: examples of clusters (second column), nature (third column) and typical binding energies (last column). After [14].

Type	Examples	Nature of binding	Binding energy
Ionic clusters	$(NaCl)_n$, Na_nF_{n-1}	Ionic bonds Strong binding	~ 2–4 eV
Covalent clusters	C_{60}, S_n	Covalent bonding Strong binding	~ 1–4 eV
Metal clusters	Na_n, Al_n, Ag_n	Metallic bond Moderate to strong binding	~ 0.5–3 eV
Van der Waals	Rare gas clusters Ar_n, Xe_n	Polarization effects Weak binding	$\lesssim 0.3$ eV

of cluster dynamics. We take as an example the metal cluster Na_9^+ which exhibits a binding of intermediate strength. It has a radius of about $R \sim 8\,a_0$ and charge $q = 1$. The typical electric field at the surface is $E_0 \sim q^2/R^2 \sim 3.4\,\text{eV}/a_0$. This corresponds to the field strength of a laser with intensity $I \approx 10^{13-14}\,\text{W/cm}^{-2}$ (see Section 2.2.2). This means that nondestructive analysis of clusters by light must use laser intensities much below this value. In turn, laser intensities of this order or larger will directly and quickly detach electrons and most probably destroy the cluster.

1.2.3
Cluster Science

1.2.3.1 A New Science

Cluster science is a young field which emerged as an independent scientific domain only a few decades ago. Indeed it has only been since then that one has been able to produce and control free clusters, that is in the gas phase, while embedded clusters have been used in technological applications for much longer. Their proper identification and analysis, again, lived up only recently with the steady improvement of measuring tools as, for example, the scanning tunneling microscope (STM). Still, within a relatively short period of a few decades, cluster science has gathered an impressive amount of insight and has also inspired many other areas of science. There are probably few scientific domains which experienced such a rapid development. This was certainly favored by the many relations of cluster science with other fields such as atomic, molecular physics, and chemistry on the one side, and solid-state physics on the other. They deal with "similar" systems, and yet clusters are different, as we have seen above, which make comparisons interesting and fruitful. As an emerging domain, it also has the spice of interdisciplinarity. Cluster science has gathered scientists with backgrounds from several longer established domains, all bringing their own, specific scientific and technological knowledge, thus creating a marvelous learning environment for all of them.

Cluster science even goes beyond these close related fields, as it also finds many applications in other apparently remote scientific domains such as biology or astrophysics. This is, if needed, another beautiful justification for the interest in studying clusters, even if they may represent key systems in other domains. Looking at those remote fields inspires cluster physics once more, as a complete understanding of clusters, for example, for astrophysical or biological applications, brings up new questions for cluster research.

These few general ideas lay out the guidelines we have tried to follow throughout this book. We develop them a bit further in this last section before attacking more specific aspects in the next chapters.

1.2.3.2 A Quantum Mechanical Playground

Quantum mechanics is crucial for understanding cluster properties as it is in molecular, solid-state physics, or chemistry. In this respect, cluster science might just appear as another playground for quantum mechanics. This would overlook a key aspect, which we already outlined in Section 1.2.1, namely scalability. Clusters are one of the few, if not the sole, system which can interpolate between atom and bulk, and which can thus provide a unique class of systems for understanding how matter is formed of atoms.

As is well known, the quantum many-body problem is very demanding and has known solutions only for rather a few cases. The systems with very few constituents (the helium atom or a few dimer molecules with one or two electrons, ...) allow highly elaborate approaches and can be treated almost exactly. On the other extreme, bulk matter with a lot of symmetries again permits very detailed computations as, for example, for the homogeneous electron gas if the ionic background is simplified to homogeneous jellium. These (almost) exact solutions play an essential role in the many-body problem as benchmarks of the test of approximations and inspiration for the development of further approximations. In between these two extremes, in terms of size and symmetry, one is bound to develop approximate and more or less elaborate theoretical schemes. This holds for molecules as well as for solids. Clusters interpolating between the two extremes allow, in particular, to test modeling continuously at all scales.

There is a further interesting aspect in relation to the quantum many-body problem. Clusters belong to the large class of finite fermionic systems, in which one can include atoms, molecules, quantum dots and nuclei. As we shall see in Section 6.1, there are indeed strong similarities between all these systems which as such are useful to collect. Moreover, they can often be treated with similar approaches and methods. Comparing the performance of these methods in the various fields helps to find out the limitations and to subsequently improve the methods.

1.2.3.3 Interactions with Related Fields

As mentioned above, clusters find applications in many domains of science, in obviously closely related fields as well as in apparently more remote areas. We will address selected examples in Section 5.3 and Chapter 6. Here, we briefly summarize and preview some examples.

Cluster science has direct and "natural" connections with material science. The developing capabilities to manipulate materials at the nanoscale is a typical example. The many principle studies led on embedded or deposited clusters (see, for example, [15]) may for example find direct applications in dedicated shaping of nanodevices, see, for example, [16]. Small Au clusters on surfaces serve as efficient catalysts, metal clusters are considered as nanojunctions in electrical circuitry, and the coupling to light is exploited in producing an enhanced photocurrent by depositing Au clusters on a semiconductor surface.

There are also many applications of clusters science in biology and medicine (see Section 6.4). One typical example is the use of coated metal clusters in treatment and medical imaging, thanks to their optical properties and their remarkable ability to easily couple to an external electromagnetic field.

More surprising are astrophysical applications. The understanding of the composition of the interstellar medium is a key issue in astrophysics, for example, for the transmission of electromagnetic signals emitted by stars. It turns out that the interstellar medium contains a sizable amount of carbon structures, known as polycyclic aromatic hydrocarbon (PAH) ones which might play a key role in light absorption. The study of this class of clusters has become an important issue in astrophysics over the past few years, see Section 6.2.

Clusters are also of great importance in climate issues since aerosol particles, that is submicron particles, turn to play a key role in the formation of clouds, and also in the modification of their reflecting properties. Therefore a deeper understanding of the formation and growth of such nm objects will certainly help to better predict the radiative balance of the Earth's atmosphere. Key aspects will be addressed in Section 6.3.

Looking back at these examples demonstrates, again, the double use of cluster physics. On the one hand, there is a great variety of clusters important for applications. On the other hand, clusters are interesting as objects *per se* and provide ample testing ground for basic physics.

2
Measuring Clusters

The analysis of cluster properties requires the understanding of how clusters can be produced and what quantities can be measured. It is the aim of this chapter to address both these aspects. Differing with other systems, such as atoms or atomic nuclei, most free clusters are not readily available but have to be synthesized in several stages of an experimental setup. This sets some limitations on the "quality" of the produced clusters, as we shall see. The first part of this chapter presents in brief the most important cluster production mechanisms and cluster sources. We will address here the limitations set by production techniques concerning cluster size, charge, and temperature. The second part discusses the handling of clusters and the measurement of their properties. It reports basic experimental investigations on clusters in terms of experimental tools and delivered signals.

Cluster production and analysis are intimately linked. A typical experimental setup integrates all elements from the source to the collection of data. This is made possible by the relative compactness of all the elements (production/measurements/data storage) in the experimental setup. Cluster physics experiments can be performed in rooms of a few tens of square meters at most. Figure 2.1 provides a schematic picture of a typical experimental setup. It is constituted of three stages: production, perturbation and detection. The three stages are schematically illustrated, showing for production and measurement the most widely used devices. Production is represented by a supersonic jet source (see Section 2.1.1). Perturbation is an electromagnetic pulse as delivered by a laser or charged particles (ions or electrons). Measurement is illustrated here by a time-of-flight (TOF) spectrometer, one of the most widely used mass spectrometers in cluster science (see Section 2.3.1). Let us briefly discuss this setup in a few words. Once produced in the source (Figure 2.1a), clusters are ionized either by an electron/ion or photon impact. The nonzero net charge thus acquired is essential as it allows mass identification in the TOF mass spectrometer. As we shall see below, identifying cluster mass indeed turns out to be a key issue. Figure 2.1 thus provides a minimum experimental setup in cluster science. As is obvious, a TOF measurement gives access only to cluster mass (actually mass over charge ratio). In many cases, this is the aim of the measurement, as we will see in Section 2.3.1. But in most cases, it is only the first stage in which clusters with well-defined mass and/or charge are provided. Further detection devices then follow, for example, to measure properties of

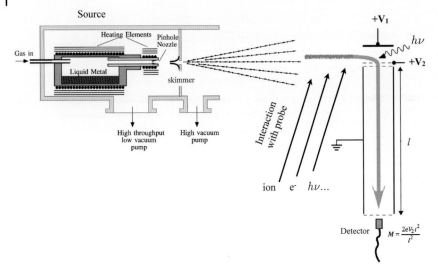

Figure 2.1 Schematic view of an experimental setup. The left lower part sketches a production device, the widely used supersonic source (Section 2.1.1). The scheme shows its basic elements: furnace (left), injection of inert gas carrier (extreme left), heating system (center), nozzle (right) and skimmer (extreme right) ahead of an expansion vessel. The middle part indicates the actual perturbation of the system, by interaction with a laser delivering photons or a charged projectile (electron or ion). Finally, the properties of the irradiated species are analyzed in a detector which is sketched in the right part. We have chosen here to schematically represent the widely used time-of-flight (TOF) spectrometer (see Section 2.3.1).

emitted electrons (see Section 2.3.4). Not shown in the sketch is the very last step of the chain, namely data storage and analysis, the latter usually done offline. In any case, it should be clear that the overall experimental chain runs in one stroke through from production up to data storage. Because the setup is usually operated as a whole, the cluster production phase is of particular importance as it directly affects experimental outcomes. One can thus say that source management appears as an integrated step in the measurement process itself. This is the reason why we shall specifically discuss it in Section 2.1.

2.1
Cluster Production

As already pointed out, the ability to produce clusters with well-defined properties has been the prerequisite for the booming of cluster physics. Cluster science, in particular the study of free clusters, really started in the 1970s, thanks to the availability of sources for free clusters. Lacking controlled means of production, most of the previous studies were performed on embedded or deposited systems. Although such cases are relevant for most applications and highly interesting as such (see Chapter 5), it is often hard to disentangle the origin of observed phenomena in the complex mixing of the effects due to the cluster itself, substrate or matrix. The study of free clusters is thus directly related to progress made at the side of cluster sources. Not surprisingly, many types of sources have been developed over the years to produce clusters of various types and sizes. There are essentially two basic mechanisms to produce clusters, which are finite objects or finite pieces of material. One can either go upward or downward in size. This means that one can either aggregate smaller systems (atoms, molecules, small clusters) or break larger systems (bulk typically). And some cluster sources follow a mixed strategy.

We will briefly address the involved production mechanisms in more detail. In the following, we shall very quickly present the various types of frequently used cluster sources, that is supersonic jets, aggregation sources and surfaces sources. We will pay special attention to the example of the widely used supersonic jet sources which we also use to exemplify the general experimental difficulties raised by cluster production. We shall then discuss the problems of cluster identification and properties such as size, charge, and temperature in Section 2.1.2. More detailed discussions of various sources and their advantages and disadvantages can be found in the review [17] or in the books [18, 19].

2.1.1
Cluster Sources

2.1.1.1 Supersonic Jet Cluster Sources

Supersonic jet sources are widely used and their functioning is rather well understood. We thus take them as a prototypical example of cluster source relying on cluster formation by condensation of an expanding gas of atoms [20]. This is achieved by compressing an inert gas (typical total pressure $P \sim 10$ bar) initially seeded with atoms of the material one wants to make clusters of. This compressed seeded gas is then squeezed through a small nozzle (Figure 2.1a for a schematic picture). The ensuing adiabatic expansion and the associated cooling slow down the atoms up to a point at which binding between neighboring atoms becomes energetically favorable, thus starting condensation. This then leads to the successive aggregation of the atoms into clusters. The carrier gas, being inert, does not participate in the binding. It only serves to provide proper building conditions in terms of density and temperature.

Supersonic sources are often used for producing metal clusters of low melting point (typically alkali metals). A furnace contains the molten metal which is heated to produce a metal vapor of (low) pressure around 10–100 mbar. This vapor is then mixed with (that is seeded into) the high-pressure rare gas. The hot mixture of metal vapor and rare gas is driven through the nozzle, expands and cools down after the passage through the nozzle. Seeded supersonic nozzle sources are mostly used to produce intense, cold, and "directed" cluster beams with acceptably narrow speed distributions. A major advantage of this kind of source is that the dynamics of the expansion allows to control the condensation process. They permit the formation of clusters with hundreds, even thousands of atoms per cluster in reasonable abundance. Let us now briefly discuss the underlying microscopic mechanisms.

One starts with an initial gas of atoms at a given temperature T_0. Before expansion through the nozzle, atomic velocities are random. The initial high pressure implies a small mean free path (much smaller than the nozzle diameter) and thus a highly collisional regime. This regime is typically hydrodynamical. The adiabatic expansion through the nozzle leads to a strong "alignment" of the atomic velocities because the expansion velocity is much larger than the original thermal atomic velocities. All in all, the first phase of the expansion basically remains hydrodynamical, even if temperature, pressure and density of the gas are strongly reduced. It is precisely during this expansion phase that a proper "window" of density and temperature opens up, allowing clusterization (or condensation) to occur. As expansion proceeds, one finally reaches a point beyond which the picture of a continuous medium breaks down, and so does the hydrodynamical approach. From then on, each created cluster will more or less follow its own path. The basic properties of the further expanding gas are essentially frozen beyond that "decoupling" stage. This is for example the case of the temperature of the system, which, as we will see below, may have a significant impact on measured quantities (see Section 2.1.2).

The expansion leads to cluster formation via condensation. This process is complex and not yet understood on a fully quantitative basis, although the basic condensation mechanism is simple. Atoms can bind to form a dimer when temperature in the jet sufficiently drops down, that is when temperature becomes smaller than the binding energy of the dimer. These dimers (some being possibly already present initially) constitute seeds for further clusterization. It is interesting to note that the kinematics of the expansion favors condensation, because it leads to a small spread of atomic velocities, thus keeping atoms in the vicinity of each other. The further evolution of the system towards the formation of large or small clusters depends on the thermodynamical properties of the jet itself. For low jet pressures, cluster growth mostly proceeds on the basis of monomer aggregation which preferably produces low-mass clusters. Higher jet pressure P, in turn, allows growth of clusters not only by atomic but also by cluster–cluster aggregation. The net result is the production of larger clusters. The initial pressure P_0 of the atomic gas thus plays a key role in determining the actual size distribution of the produced clusters and thus the abundance spectrum (see Figure 2.2).

This brief overview of cluster production gives an idea of the difficulties one may face in identification and characterization of the formed species. It is clear that the proper tuning of source parameters is an essential ingredient. There are unfortunately no definite models to predict these parameters precisely, even if basic underlying physical mechanisms are clearly identified. As guidelines, empirical scaling laws exist, for example the Hagena parameter Γ^*, used in the formation of rare gas clusters. It reads $\Gamma^* = \kappa P_0 D^q T_0^{0.25q-2.25}$ where P_0 and T_0 are the initial pressure (expressed in mbar) and temperature (in K) of the atomic gas, respectively, D (in µm) is the diameter of the nozzle of the supersonic source, and q is a parameter typically between 0.5 and 1. One usually takes $q = 0.85$, while κ depends on the rare gas. For instance, one uses the value 5500 for Xe, 2890 for Kr, and 1650 for Ar.

Γ^* turns out to be also proportional to the average number of atoms per cluster. Therefore cluster size increases with increasing P_0 or D, and decreasing T_0. Finally, impurities in the gas play a role by constituting condensation germs which usually lead to an increase of average cluster sizes. After all, proper handling of a source remains a matter of intuition and experience. The distribution of formed clusters is usually broad (see Figure 2.2) and requires further selection (see Section 2.1.2.1).

2.1.1.2 More on Cluster Sources

Let us now consider gas aggregation sources which provide a simple and efficient means to produce large clusters and in which the basic production mechanism is again condensation, as in supersonic jets. These sources are laboratory copies of a smoking fire or of cloud and fog formation in nature (see Sections 6.3.3 and 6.3.4 on the formation of clusters in a climate context). More precisely, a liquid or a solid is evaporated into a colder gas: it cools down the evaporated atoms or molecules until condensation starts. Condensation then roughly proceeds as in supersonic jets where here the dominant clustering mechanism is successive monomer addition. Different from supersonic jets, here atoms are injected more gently in a stationary or streaming gas. Thus the condensation is not ruled by expansion dynamics but is closer to thermal equilibrium processes which hinders the natural stopping of cluster growth by the geometry of the expansion. Although the condensation process is here more complex and harder to control, and in spite of lower delivered beam intensities, gas aggregation sources have been quite successful. The most famous example of such (so-called) smoke sources are the facilities used to produce fullerenes (C_{60}, C_{70}, ...), but these sources have also been used to produce metal clusters [17].

Surface sources rely on the basically different production principle by break up from a larger compound. The idea is to remove finite pieces of material from a solid surface by energy impact (photons, electrons, ions). This ablation phase is usually complemented by a further clustering (condensation) phase. In surface erosion or sputtering sources, ablation is attained by impact of a heavy (charged) particle which ejects atoms, molecules or clusters from the surface. Ablation is achieved by photon impact in a laser evaporation source (LES), and by a high-current pulsed arc discharge in pulse arc cluster ion source (PACIS). It has also been recently shown experimentally that fullerenes can also form from graphene sheets bombarded by an energetic electron beam. In general, the handling of such sources is technically demanding. The violent initial formation by break up leads to high-temperature clusters. A way to overcome this disadvantage is to couple them to supersonic jets or aggregation sources to cool down the formed clusters. Finally, pick up sources are used to produce mixed clusters composed of various materials. The idea here is to mix clusters as formed in supersonic jets with a jet of other molecules. It is to be noted that pick up sources are also used to embed molecules or clusters in He droplets thus preparing embedded systems (see next paragraph).

2.1.1.3 Embedded and Deposited Clusters

We have up to now discussed only production of free clusters. But deposited/ embedded clusters also constitute a very rich field, often allowing studies not easily possible with free clusters (see Section 5.3). In terms of cluster growth, the substrate provides a longer time scale, thus possibly leading to the formation of larger species. Substrates also allow to attain larger densities of clusters, therefore allowing measurements with weak signals, as for example second-harmonic generation or Raman spectroscopy. Last but not least, it is also much easier to control cluster temperature via the substrate. They also offer better means for structure analysis, for example, by scanning them with STM, see as an example Figure 1.1b.

The first method to produce embedded/deposited cluster is growth from a supply of atoms or ions. An example is the growth of metal clusters on an insulating substrate by exposing the substrate to a metal vapor. Another example in the case of embedded clusters is the growth of metal clusters in glass by diffusion from a surrounding molten metallic salt. Cluster size can then be more or less controlled by tuning pressure and time. But such methods are applicable only if the combination of materials tends to clustering of the vapor atoms or ions, respectively. Note also that deposited clusters may have a high mobility on the surface, which may perturb long time measurements by drift and reactions.

The alternative method for production of embedded or deposited clusters consists of first producing free clusters and then depositing them onto a substrate or into a matrix. The process strongly depends on the velocity of clusters. Deposit with low-energy impact places the clusters more or less gently on the substrate, leaving the clusters mobile and thus allowing rearrangements by diffusion along the surface and reactions. The medium energy impact case pins small defects into the surface which fixes the deposited clusters, while the high-energy impact deposit causes severe damage producing a new composite rather than a deposited cluster. Low-energy deposit is often still too hefty. Even more gentle deposit (soft landing) can be arranged by covering the substrate with a few rare gas layers on which the impinging clusters are safely stopped. If enough clusters are collected, one evaporates the rare gas and remains with almost unperturbed clusters deposited on the substrate.

Mass selection is usually a problem with embedded/deposited clusters. The growth from vapor produces a broad distribution of masses and shapes, only roughly controllable through the growth conditions. The deposit of free clusters allows, in principle, the use of originally mass selected clusters. But the actual attachment possibly changes the clusters' shape and produces again some spread in masses. There are possibilities to improve *a posteriori* the mass distribution by selective destruction of unwanted species through intense laser fields. Another problem concerns the theoretical description of embedded and deposited clusters. As the environment has a significant impact on cluster properties, models have to deal with both cluster and substrate, which is obviously more demanding than considering a single free cluster. Efficient theoretical methods are still in a developing stage, see Section 5.2. On the other hand, the many conceivable combinations of cluster/environment generate a huge variety of extremely interesting scenarios with high application potentialities. This variety leaves ample space for further research and development.

2.1.2
Sizes and Temperatures

The above brief overview of the most commonly used types of cluster sources demonstrates how cluster production is a delicate but an essential step of any experimental program. Cluster sources usually leave us with poorly known cluster properties as size and/or temperature. These difficulties can be somewhat handled by a proper tuning of source parameters. But the basic size/temperature problem remains central, as it may strongly affect experimental findings and it also renders comparisons with theory less direct. Good measurements will require special control of both parameters, size and temperature, in order to avoid mixing results in a too incoherent manner.

2.1.2.1 Cluster Size

The simplest way to point out the size problem is to analyze the various sizes of the clusters formed by the source. Depending on the goal of the experiment, one may then have to trigger cluster sizes, for example by using a mass spectrometer. We shall present a few examples of such mass spectrometers later in Section 2.3.1.

The variations of the cluster mass distribution with physical parameters in the cluster source is illustrated in Figure 2.2. It shows the mass spectra of CO_2 clusters as obtained under various pressures, and thus clustering conditions. The initial pressure P_0 (namely inside the source itself) is varied between about 0.7 and 3 bar. In order to vary only one parameter at a time, the initial temperature T_0 of the atom gas is kept fixed at 225 K. It is obvious that the mass distribution strongly depends on P_0. Under weak clustering conditions, namely at low pressure (Figure 2.2a), a roughly exponential mass spectrum is obtained. Higher pressures (Figure 2.2b,c) leads to better clustering conditions: the higher density of raw material enhances cluster growth through cluster–cluster collisions. This finally leads to a sizable, although very broad, peak around a finite value of the average cluster size. As expected, one obtains in each case a distribution of cluster sizes. The average size increases with increasing pressure P_0. It is also noticeable that, with increasing pressure P_0, the width of the mass distribution significantly increases as well, staying comparable to the average cluster size itself (take, for example, the case at 3000 mbar (Figure 2.2c), with a peak around size 500 and a width of order 700 mass units).

Figure 2.2 exemplifies a basic difficulty in cluster production: one obtains a very broad distribution of clusters. Since clusters of different sizes exhibit different physical properties, as we will see in several areas of this book, the consequence is that one cannot usually exploit the cluster beam as such. A postprocessing is compulsory in order to mass select the clusters on which a measurement is to be

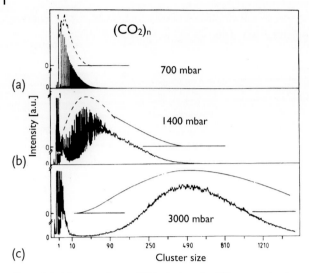

Figure 2.2 Mass spectrum of CO_2 clusters for different production pressures: 700 mbar (a), 1400 mbar (b) and 3000 mbar (c). Solid and dashed lines represent theoretical estimates of mass spectra. After [14].

performed. Therefore, the experimental resolution directly depends on the quality of identification of the clusters under consideration.

The task of mass selection raises the related question of cluster charge. Indeed, mass spectrometers basically select a charge to mass ratio "q/m" (see Section 2.3.1) and thus allow to deal only with charged species, and not neutral ones. However cluster sources "naturally" produce neutral clusters. Finer cluster selection thus requires extra charging. This can be rather easily attained by controlled laser irradiation. The laser delivers a well-tuned cluster charge state such that the q/m ratio directly defines masses in the end. So far it implies dealing with charged species. The question of studying neutral clusters remains open. There are several ways to attack this question. An obvious one consists of passing through a stage of charged clusters. The neutral cluster beam is thus first irradiated by a laser delivering charged species which can be mass selected by a standard mass spectrometer. The selected clusters are then re-neutralized by collision with an electron beam, which finally delivers a mass selected beam of neutral clusters.

Besides the advantages in handling, charged clusters as such are extremely interesting objects of study and we shall consider them at many places along this book. There are two classes: anions (negatively charged) and cations (positively charged). The former can be obtained (again starting from neutral clusters out of a source) by collision of the cluster beam with an electron beam. Clusters attach some electrons and thus become anions. Cations are easily obtained by laser irradiation which strips some electrons from the target neutral clusters. It should be noted that charging clusters involves a few charges at most. A large charge leads to Coulomb instability (the charge threshold depending on the cluster mass). For example, in Na_n, only clusters with more than 27 atoms may sustain two positive charges, and similarly only $Na_{n>65}{}^{3+}$, $Na_{n>123}{}^{4+}$, ..., are stable [21]. A large negative charge is limited by electronic binding which decreases with each added electron and becomes quickly very small. For example, the ionization potential of $Na_n{}^{2-}$ clusters (at temperature $T = 0$) vanishes around $n = 35$, and double anions with smaller n are not stable.

2.1.2.2 Cluster Temperature

Clusters are formed at finite temperature. The whole production mechanism, as discussed above, involves temperature: the cooling down of the expanding vapor leads to clustering which occurs at different stages (and thus different temperatures) of the expansion. The actual cluster temperatures results from a balance between globally decreasing temperature of the collectively expanding stream and the formation heat of the aggregation. This even makes the temperature (in detail) specific to the history of each cluster. This holds even more true as the thus formed "hot" clusters may further cool down, for example, by collisions inside the jet or by monomer evaporation. That level of detail is nevertheless hard to track and one will thus finally consider a population of clusters formed in the jet with a temperature distribution.

The question is then how cluster temperature may affect cluster properties. A better temperature control is actually highly desirable. This can be achieved by temporarily mixing the cluster jet with an inert-gas beam of well-defined temperature [22], which acts as a thermostat (see pick up sources in Section 2.1.1). This allows temperature tuning and possibly cooling down, when necessary.

In order to illustrate the importance of controlling cluster temperatures, we present in Figure 2.3 the impact of (in that case well-controlled) temperature on a key observable for metal clusters, namely their optical response (see Section 4.2). We note that the optical response, especially for metal clusters, is related to the cluster's color. The typical frequencies lies in the visible domain (ranging from 1.8 up to 3.1 eV) for simple metal clusters, as shown in Figures 2.3a–c. One observes the strong influence of temperature on the distribution. This is easy to understand. Typical temperatures are of the order of room temperature (~ 300 K) at most. But such temperatures leave electrons totally unaffected because 300 K $\sim$ 0.03 eV is negligible compared to typical Fermi energies of order a few eV. Electrons are thus practically at zero temperature. Ions, on the other hand, are strongly affected by such temperatures and thus sustain sizable thermal excitation. This means that ions will significantly fluctuate at ionic time scales. However the oscillations of excited electrons are typically two orders of magnitude quicker than ionic motion. For then, electrons have more than enough time to oscillate several times before ions noticeably move. Therefore they explore ionic configurations nearly instantaneously. Now the electronic optical response sensitively depends on the cluster shape (see Section 4.2.2). The finally measured optical response thus piles up the individual optical responses of the thermal ensemble of cluster shapes, each sample of which contributing its own specific optical response. The net effect of

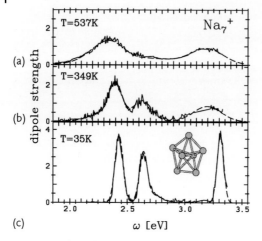

Figure 2.3 Effect of cluster temperature T on an electronic observable, here the dipole strength (that is the optical absorption strength), for 537 K (a), 349 K (b) and 35 K (c). The test case is Na_7^+ whose structure is sketched in (c). Data taken from [22].

increasing temperature is a substantial broadening of the optical peak, as is clearly visible in Figures 2.3a–c.

2.2
Excitations of a Cluster

We have seen above that the production of cluster beams with well-defined properties is quite demanding. From now on, we assume that these parameters are under control. Next comes the task of measuring cluster properties. This usually proceeds in two steps: excitation and recording emerging signals. Figure 2.4 sketches a couple of pathways to excitation and measurement. These will be explained in the following. We will address excitation mechanisms in this section and the analysis of emerging signals in the next.

Nearly all excitation mechanisms go through the electromagnetic interaction which couples directly and efficiently to the cluster electrons and ions. Some "excitations" are static perturbations in terms of an external electric or magnetic field. Most excitations, however, are dynamic as the field of a laser pulse, a bypassing ion or an electron beam. Static perturbations are usually closely related to the measured observable. A static electric field will give access to static polarizability, and a static magnetic field will allow to measure magnetic moments. In both cases, the analysis is nondestructive. One is thus directly using the cluster as an input for detection, and the system returns to its initial state if the measurement is over. Time-dependent electromagnetic perturbations offer a much broader variety of scenarios which we will outline in the following.

2.2 Excitations of a Cluster

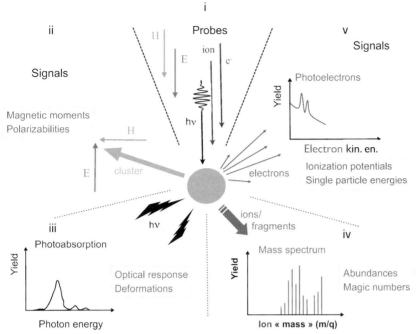

Figure 2.4 Schematic view of analysis of cluster properties. Indicated are typical excitation mechanisms, namely impact by electrons, ions, photons (lasers) or static electric or magnetic field (i). Around that are grouped various signals (clusters themselves, photons, ions and/or cluster fragments, electrons), which emerge from the excitation and which are used for quantitative analysis (ii → v). The various groups are constituted in terms of typical signals (electrons, photons, ions, ...) and associated measured quantities. Starting from (ii), we mention the measurements of electric and magnetic polarizabilities (see also Figure 2.6), themselves attained by further probing the already excited cluster by another electric or magnetic field. Optical signals themselves are indicated (iii), giving access for example to cluster shape (see Figure 2.7). Ionic signals are referred to (iv), as obtained by recording ions and/or cluster fragments, typically providing mass spectra which give in particular access to cluster abundances (see Figure 2.5). Finally, electronic signals are illustrated (v) as attained by analyzing kinetic energies and angular distributions of emitted electrons (see also Figures 2.9 and 2.10).

2.2.1
Collisions with Projectiles

Collisions with atoms, molecules, or highly charged particles are a standard means for the excitation of atoms and molecules (see for example [23, 24]) but also of clusters. Conceptually simplest are collisions with charged particles. Highly charged ions and protons can be considered as being structureless in such situations. The ions can be treated with classical trajectories $R_{ext}(t)$. For fast ions, one may approximate these trajectories by straight lines. Only their Coulomb field couples to the cluster. The effect of the ion (of charge Z_{ext}) on the cluster can be described by the time-dependent external field

$$U_{\text{ext}}(\mathbf{r}, t) = \frac{Z_{\text{ext}} e^2}{|\mathbf{r} - \mathbf{R}_{\text{ext}}(t)|} \cdot \quad (2.1)$$

Magnetic and/or relativistic effects may play a role for very fast ions, but are neglected here for simplicity. The ionic trajectories are characterized by the initial ion velocity and the impact parameter b (which is the distance of closest approach when considering a straight-line continuation of the incoming trajectory). The velocity is usually well defined by the experimental setup. But the impinging ion beam covers a broad range of b. From the theoretical side, one has to run a couple of calculations with systematically varied impact parameters and cluster orientations. Reaction cross-sections are then computed by integration of the reaction probability over impact parameters.

Electron collisions require a full quantum mechanical treatment with standard methods of scattering theory [25]. Typically, one describes the process with the distorted-wave Born approximation (DWBA), where the projectile electron explores the scattering potential in all detail (by techniques of phase analysis). DWBA approximates the cluster as an inert scattering potential. That is not applicable for low-energy electrons and for inelastic scattering. A more involved coupled channel calculation is then required.

More involved is the case of collisions with atoms, molecules or other clusters. Projectile and target have now an intrinsic electronic structure which can be relevant for the reaction. One should then employ the whole machinery of electronic and ionic propagation for both systems in equal manner. The task to compute reaction cross-sections from a bunch of trajectories with different impact parameters is the same as for collisions with charged particles.

2.2.2
Laser Fields

Lasers are the most important and very flexible means for a dedicated, well-tuned excitation of electronic systems. They produce a strong coherent electromagnetic field which can be well approximated in most cases by a classical time-dependent electromagnetic field $\propto \exp(i\mathbf{k} \cdot \mathbf{r} - i\omega t)$, times a temporal profile function $f(t)$. Typical wavelengths are in the range of several hundreds nm. This is a huge distance as compared to atoms, molecules, and (most) clusters. As an example, blue light (a short wavelength in the visible) at a frequency $\approx 3\,\text{eV}$ has a wavelength $\approx 300\,\text{nm}$, while Na clusters have a radius $\approx N^{1/3} r_s$ with the Wigner–Seitz radius $r_s = 4\, a_0 = 0.2\,\text{nm}$ [10]. It requires a cluster as large as $N \approx 1.7 \times 10^9$ until the diameter equals the wavelength. Actual cluster sizes usually stay far below that value. One can thus treat the laser field in the limit of long wavelengths ($k \to 0$), that is we deal with a spatially homogeneous electrical field $\mathbf{E}$

$$\mathbf{E}_0 \exp(i\mathbf{k} \cdot \mathbf{r} - i\omega_{\text{las}} t) f(t) \to \mathbf{E}_0 \exp(-i\omega_{\text{las}} t) f(t), \quad (2.2a)$$

and we can neglect the effect of the magnetic field. This laser field is characterized by the peak field strength $E_0 = |\mathbf{E}_0|$, the polarization $\mathbf{e}_\mathrm{pol} = \mathbf{E}_0/E_0$, the time profile $f(t)$ of the pulse, and the frequency ω_las. The corresponding photon energy is thus $\hbar\omega_\mathrm{las}$, usually expressed in eV (see list of units at the beginning of the book). By a misuse of language, we will also express laser frequencies in eV throughout the book. The laser field strength is usually given in terms of the laser intensity I as

$$E_0 = c_{EI}\, I^{1/2}\,, \quad c_{EI} = 1.07 \times 10^{-8}\, \frac{\mathrm{eV}}{a_0}\left(\frac{\mathrm{W}}{\mathrm{cm}^2}\right)^{-1/2}. \tag{2.2b}$$

The profile is a matter of debate. Experimental pulse profiles are not precisely known. They are assumed to be a well-peaked function with a certain full width at half maximum (FWHM), often well approximated by a Gaussian, although one has to be aware of a possible pre-pulse background in case of short pulses. The practical difficulty in computations is that the Gaussian signal spreads over all time. Theoretical applications prefer pulse shapes with definite switching times. A good compromise, still close to the experimental shape and still having high spectral selectivity, is the $\sin^2$ pulse envelope

$$f(t) = \theta(t)\,\theta(T_\mathrm{pulse} - t)\, \sin^2\left(\pi\frac{t}{2T_\mathrm{pulse}}\right). \tag{2.2c}$$

The form (2.2c) is scaled such that the pulse parameter T_pulse is identical with the FWHM. So far, Eqs. (2.2a)–(2.2c) model a single pulse. Modern experiments go beyond that. Typical pump-probe experiments employ a succession of two pulses with a well-defined time difference [26]. Some setups tailor the pulse with a dedicated chirp (small variation of frequency during the pulse) in order to induce wanted chemical reactions. The variability of lasers is so huge that one cannot discuss all conceivable profiles on general grounds. Each case has its own very specific features and needs to be separately discussed.

The coupling Hamiltonian is formulated in terms of the electromagnetic potentials of the photon field. These are not unique due to the freedom of gauge transformation. One way is to express the electrical field through a vector potential $\mathbf{A}(\mathbf{r}, t)$ such that $\mathbf{E} \propto \partial_t \mathbf{A}$. The vector potential is spatially homogeneous in the limit of long wavelengths. The coupling $e(\mathbf{A}\hat{\mathbf{p}} + \hat{\mathbf{p}}\mathbf{A})$ becomes in detail

$$U_\mathrm{ext} = e\,\mathbf{E}_0\, F(t)\hat{\mathbf{p}}\,, \quad F(t) = \int_0^t dt'\, f(t')\exp(-i\omega_\mathrm{las} t'). \tag{2.3}$$

This form is called the velocity gauge as it employs the electron velocity via the momentum $\hat{\mathbf{p}}$. Alternatively, one can express the laser field in terms of the Coulomb potential Φ such that $\mathbf{E} \propto -\nabla\Phi$. The coupling Hamiltonian then reads

$$U_\mathrm{ext} = e\,\mathbf{E}_0\, f(t)\cdot \hat{\mathbf{r}}\,\exp(-i\omega_\mathrm{las} t). \tag{2.4}$$

This is called the space gauge because the laser field acts here simply as a time-dependent spatial dipole operator. Both formulations are connected by the well-known gauge transformation of fields and wave functions [27]. It is a matter of practical considerations which formulation one prefers. The coupling in space gauge

(2.4) is much simpler to implement. It is the form of choice for most applications. But the growth $\propto r$ can cause trouble in large numerical boxes and highly excited situations. Most observables are independent of gauge.

After all, a few words are in order about what kind of perturbation a laser exerts on a cluster. We exemplify this by using optical lasers, with photon energies of order $\hbar\omega \sim 3$ eV. Such an energy is close to the range of typical cluster energies (see Table 1.2) and in some cases, it may suffice to directly ionize the clusters. There remains a question whether the effect is purely energetic or such a laser pulse also affects electronic or ionic momenta. We consider the effect on a metal cluster, where electrons can to a large extent be viewed as a free Fermi gas [28] characterized by its Fermi momentum $p_F = \hbar k_F \sim 0.3$–0.4 eV fs a_0^{-1}. The laser photon momentum $p_{las} = \hbar\omega/c \sim 5 \times 10^{-4}$ eV fs a_0^{-1} is negligible as compared to the electron momentum. This estimate clearly shows that the impact of the laser on electrons is purely energetic and does not contribute to their momenta. The impact on ions is even smaller. One can estimate typical ionic momenta from ionic vibration energies $E_{vib} \approx 10$ meV, which leads to momenta of order $p \sim \sqrt{2M\,E_{vib}} \sim 4$ eV fs a_0^{-1} for Na, that is one order of magnitude larger than typical electron momenta and thus even farther above photon momenta. Similar relations are found in other types of clusters. The conclusion that photons bring in energy but no momentum holds in general.

2.2.3
Coupling to Light and Optical Response

Coupling to light plays a large role in cluster physics. This particularly holds for metal clusters whose surface plasmon (see below) couples extremely efficiently to light. In any case, the strong and almost immediate coupling of the electrons to an electromagnetic perturbation, especially to laser pulses, is a key mechanism to cluster measurements. We illustrate the point in this section on the case of metal clusters for which the mechanism is especially transparent. To that end, we take up the simple view introduced by Mie [3] which, in spite of its simplicity, provides a pertinent description of the gross features of optical response in metal clusters.

There are several ways to derive the Mie plasmon frequency. Often used is the model of a metal cluster being a small dielectric sphere with huge conductivity [12]. We consider here a more microscopic picture of an electron cloud oscillating against the positive ionic background. The ionic background is simplified to a homogeneous positively charged sphere of radius R and density ρ_0, that is $\rho_{jel}(r) = \rho_0 \theta(R-r)$. Electrons are modeled as a negatively charged sphere with the same density ρ, ensuring overall neutrality of the system in the ground state. We furthermore assume a collective response of the system in that the electrons move as a rigid sphere against the ions (their much larger mass allows to consider them as immobile). This assumption of collectivity turns out to qualitatively reproduce the actual response of a metal cluster. The computation of the Mie frequency starts from the energy of the displaced electron cloud in the Coulomb field of the ions

$$E(d) = \int d^3r\, \rho_{el}(|\boldsymbol{r} - d\boldsymbol{e}_x|)\, U_{ion}(r)\,, \quad \Delta U_{ion} = -4\pi e^2 \rho_{ion}\,,$$

where d is the displacement, $\boldsymbol{e}_x$ its direction, and U_{ion} is determined from ρ_{ion} by the Poisson equation. The actual direction of displacement is unimportant as we deal here with spherically symmetric U_{ion} and ρ_{el}. We consider the energy in the small amplitude limit up to order d^2. The curvature at $d = 0$ becomes

$$\begin{aligned}
\partial_d^2 E &= \int d^3r\, U_{ion}(r) \left(\partial_d^2 \rho_{el}(|\boldsymbol{r} - d\boldsymbol{e}_x|)\right)\bigg|_{d=0} = \int d^3r\, U_{ion}(r) \nabla_x^2 \rho_{el}(r) \\
&= \frac{1}{3} \int d^3r\, U_{ion}(r) \Delta \rho_{el}(r) = \frac{1}{3} \int d^3r\, \rho_{el}(r) \Delta U_{ion}(r) \\
&= -\frac{4\pi e^2}{3} \int d^3r\, \rho_{el}(r) \rho_{ion}(r) = \frac{4\pi e^2}{3} N_{el} \rho_0\,,
\end{aligned} \quad (2.5)$$

where we have used spherical symmetry to justify $\nabla_x^2 = \frac{1}{3}\Delta$ and in the last step the homogeneous distributions for ρ_{el} and ρ_{ion}. The mobile electron cloud has total mass $M = N m_{el}$ and will thus undergo harmonic oscillations with a frequency $\propto \sqrt{\partial_d^2 E / M}$, that is

$$\omega_{Mie}^2 = \frac{4\pi \rho e^2}{3 m_{el}} = \frac{e^2}{m_{el} r_s^3}\,. \quad (2.6)$$

This oscillation frequency is known as the Mie frequency.

The Mie frequency stands for a dominant eigenfrequency of the cluster which will lead to a large resonant response to a laser field in that frequency range. In simple alkaline metals, ω_{Mie} lies in the visible part of the electromagnetic spectrum, hence the term "optical response" to characterize this phenomenon. As a consequence, coupling between light and the plasmon is especially efficient. For alkalines, whose Wigner–Seitz radii r_s typically lie between 3 and 6 a_0, the Mie frequencies range between 2.1 eV (Cs) and 4.5 eV (Li), with a value of 3.4 eV for the often studied Na case [10]. These values fit quite well more refined theoretical computations and experimental data (see Sections 2.3.3, 3.4.4, and 4.2). The relations are more involved for other metals as, for example, noble metals. Here one needs to account for the polarizability of the core electrons. Doing that properly yields again a fairly good description of the dominant resonance frequency. In nonmetallic systems, the above simple Mie picture is not applicable, but a strong coupling between electrons and light remains. Then it mostly involves individual electrons rather than collective motion. The generic term of optical response applies to all sorts of photon-cluster coupling, and the term plasmon (better Mie plasmon or surface plasmon) to collective motion as experienced in metals.

2.3
Measuring Cluster Properties

Equally broad is the variety of signals used for subsequent analysis. It may be from photons, electrons, ions or clusters or pieces of clusters, depending on the nature and intensity of the perturbation.

Static perturbations apply electrical or magnetic fields and measure the response of the clusters in terms of corresponding moments (electrical dipole, magnetic moment). This determines the (static) polarizability or magnetic response.

Ionic signals have been widely exploited since the early days of cluster physics. The simplest example is an excitation by a moderate laser pulse. The cluster acquires a small net charge and can then simply be identified in a mass spectrometer. Even that seemingly trivial signal carries a lot of interesting physics. In particular, such mass spectra provide the relative abundances of cluster species and thus allow to identify especially abundant, and therefore stable, species. This analysis played a major role in the early developments of metal clusters, allowing to identify "magic" electron numbers corresponding to electronic shell closures related to enhanced stability. In the case of more violent perturbations, the cluster may be partly fragmented and emit smaller clusters and/or ions. These signals provide valuable information on the underlying dynamical scenarios.

Photons play a role in several measuring schemes. The most widely used signal is, in fact, a missing signal, namely photoabsorption. This directly measures the optical absorption strength related to the optical response of a system (see Section 2.2.3). This is a key observable giving insight into the electronic structure of a cluster and allows for metal clusters to conclude on the global shape by virtue of the dominant Mie plasmon (see Figure 2.7 and Section 4.2). Optical absorption is also the doorway to all sorts of cluster dynamics induced by intense laser pulses. Photon emission following excitation is exploited, for example, in Raman scattering. Violent cluster excitation emits photons up to the X-ray range (see Figure 2.8), whose spectra deliver information about the lattice field effects on deep lying electron states in a cluster.

Electronic signals following laser excitation can be recorded in great detail and provide plentiful information on static and dynamic cluster properties, see Section 4.3. The kinetic energies of the emitted electrons, called photoelectron spectra (PES), provide rather direct information on the single-electron energies in the cluster ground state (see Figure 2.9). PES can be further complemented by recording the angular distributions of the emitted electrons, called photoelectron angular distribution (PAD). They provide information about the spatial structure of the emitting system. Sometimes, PES and PAD are simultaneously recorded leading to a doubly energy and angle-resolved analysis of the emission yield, often denoted as "velocity map imaging" (VMI). Such measurements are very demanding as they combine a subtle analyzing tool with the by no means trivial task to produce a clean, and yet strong, cluster beam first. Nonetheless, VMI measurements are becoming more and more fashionable in cluster physics.

2.3.1
Mass Distributions

We start our brief review of cluster measurements by the simplest case of abundance spectra which, in spite of their apparent simplicity (merely a protocol of the number of clusters of a given type/size), provide invaluable information on basic cluster structure properties. One often finds some "special" clusters which are significantly more abundant than others in the sequence. And the overabundance of a given species indicates an enhanced stability coming from a particularly favorable arrangement of electrons and/or ions in the system.

The situation is somewhat similar to the case of abundances of nuclear isotopes in the universe: in our galaxy, some specific nuclei are much more abundant than their neighboring nuclei. This reflects neutron and/or proton shell closures in these specific nuclei, which explain the particular stability of these systems [29]. The very same pattern have been observed in ensembles of metal clusters, where electrons are quasi-free fermions and behave as nucleons in nuclei. In that case, an electronic shell closure also provides enhanced stability (see Section 4.1.1). However metal clusters, unlike nuclei, can be built up to an arbitrary size. This allowed for the first time to track these electronic shell closure effects, related to abundances, up to very large systems containing thousands of atoms (see Figure 4.2). However, overabundances in clusters are not necessarily to be associated to electronic shell closures only. One also observes ionic shell closures, namely special arrangements of atoms (geometric shells), which minimize surface energy and so, provide enhanced stability (see Figure 4.3). Both sorts of shell effects can appear in one and the same material. Note however that atomic shells tend to prevail at very low temperatures, while electronic shell closures in metal clusters dominate above melting temperature [30]. Particularly strong effects emerge if both mechanisms cooperate. We illustrate the case here with the very famous example of fullerenes in Figure 2.5 where C_{60} (but not only that one) exhibits a particular stability by a closed electronic shell as well as a particularly favorable geometry (see Figure 1.8d).

The typical apparatus to measure masses is the time of flight (TOF) spectrometer. It is certainly one of the simplest spectrometers. It is cheap and therefore, very much used. Still, it remains efficient and allows high levels of resolution. The TOF spectrometer is illustrated in Figure 2.1c as a final stage of the general experimental setup. The basic idea of such a spectrometer is to record the arrival time of ions, initially accelerated and then left to freely propagate before detection. Neutral clusters are first ionized ($X_N \to X_N^{q+}$) to a known charge q by a well-tuned laser field, and then accelerated by a static electrical field (capacitor) up to potential V. Having thus acquired a kinetic energy V, the ions then propagate in the field-free region with a constant velocity $v = \sqrt{2Vq/m}$, uniquely related to the q/m ratio. The simplest tuning uses low laser intensities, predominantly producing ions with $q = 1$ so that the ion velocity scales as $v = \sqrt{2V/m}$. To reach the end of the field-free region (distance D) then takes a time $\Delta t = D/v = D\sqrt{m/(2V)}$. Measuring Δt and controlling D and V thus means measuring mass m. The spatial extension D plays a crucial role for mass resolution: the larger D, the better the res-

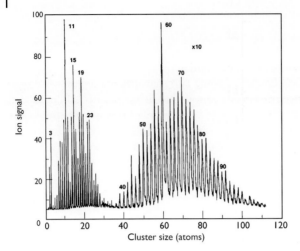

Figure 2.5 Mass spectrum of carbon clusters. They have been produced in a supersonic jet source. Fullerenes correspond to the long distribution of even-numbered carbon clusters starting around C_{40} and up to about C_{100}. The famous C_{60} appears as particularly abundant. This figure corresponds to the first published experiment revealing fullerene cluster distribution [31].

olution. Hence one often extends the TOF apparatus (see Figure 2.1) to the more sophisticated "reflectron" where the particle beam is turned by 180° at the end of the first path and recorded near the entry of the machine. This almost doubles D and it even compensates for velocity fluctuations (too fast ions take longer to revert their direction than slower ones). This finally allows resolutions $\Delta m/m$ of order 10^{-5}, about 10–100 more than in a simple TOF.

2.3.2
Magnetic Moments

Magnetic moments characterize the response of a cluster to a static magnetic field and they can be measured by a typical Stern–Gerlach device [32]. A cluster jet is injected between two magnets which generate an inhomogeneous magnetic field B. A cluster with a magnetic moment μ couples to B with potential energy $E_p = -\mu \cdot B$, and associated force $-\nabla E_p$, and torque $\mu \wedge B$. The inhomogeneity of B delivers a net force on the cluster, provoking a deflection of the cluster beam, whose amplitude depends on the magnetic moment and/or cluster size. The deflections are then recorded and a mass identification performed.

An example of a magnetic moment measurement is shown in Figure 2.6. As is well known, the magnetic moment is proportional to the total angular momentum of a system and, in the absence of a global rotation, directly proportional to the spin itself. The effect is thus larger in the magnetic metals Ni (Figure 2.6a), Co (Figure 2.6b), and Fe (Figure 2.6c). Figures 2.6a–c thus show magnetic moments per atom as a function of size for clusters made from these elements. The three

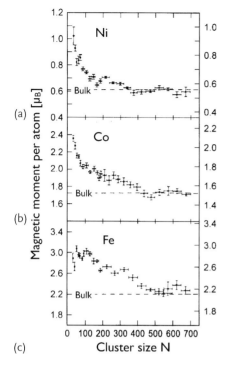

Figure 2.6 Average magnetic moments of size-selected clusters of various materials: Ni (a), Co (b) and Fe (c), in units of Bohr's magneton μ_B. The horizontal dashed line indicates the corresponding bulk value. The estimated temperature of the clusters is around 100 K. After [33].

systems differ by their behaviors in bulk (see horizontal dashed line): the bulk magnetic moment per atom is about four times larger in Fe than in Ni, and about 1.5 larger than in Co. On the other hand, the general dependence of the magnetic moment with size is similar for the three materials. All cases show a progressive decrease towards the bulk value. But the latter one is reached at different cluster sizes, depending on the material. There are furthermore oscillations of the size dependence, especially for small clusters, reflecting shell effects. The oscillation amplitudes also differ from one material to the next. It should be finally noted that temperature effects play a crucial role on magnetic moments as well. They fade away with increasing temperature. All these trends, similarities and differences reflect a typical feature: cluster properties do sensitively depend on size (and temperature). We concluded this in a similar fashion from Figure 1.7, in connection with a totally different observable, namely the electron affinity.

2.3.3
Photon Signals

The basic photon signal is photoabsorption cross-section related to the optical response of a system. It is the response of a cluster to a time-dependent electromagnetic dipole field, in most cases from a laser. The dipole approximation is applicable because clusters are usually smaller than the wavelength of (visible) light. In turn, photoabsorption is extremely selective and exclusively measures dipole modes. Its cross-section is directly proportional to the energy-weighted dipole strength or dynamic dipole polarizability, see Sections 3.4.4 and 4.2.

In general, optical spectra display a fragmented spectrum with a series of many peaks mapping the excitation spectrum of dipole of the system. These eigenmodes can either reflect a particle–hole structure or a collective nature (Mie plasmon), depending on the system and energy range. Spectra of metal clusters are distinguished by one strong resonance peak usually in the range of optical frequencies. The basic underlying mechanism is the Mie surface plasmon, viewed as a collective oscillation of the valence electron cloud (rather "soft" in metal clusters) with respect to the ionic background, as was already discussed formally in Section 2.2.3. There is also a strong impact of temperature on the optical response as was discussed in Section 2.1.2.2 and exemplified in Figure 2.3. This high sensitivity is particularly important for clusters and stems from their high density of isomers.

The collective nature of the optical response in simple metal clusters allows to draw conclusions on the underlying global cluster deformation (see Sections 4.1.2 and 4.2.2). The point is illustrated in Figure 2.7 showing three Na clusters of moderate size but with different deformation: spherical (Figure 2.7a), oblate (Figure 2.7b) and prolate (Figure 2.7c). The striking feature is the qualitative differences between the three spectra. Although the three clusters are of comparable sizes (with consequently a comparable range of energies for the Mie plasmon), the shapes of their spectra are completely different, exhibiting one or two peaks of various intensities. This behavior can be directly related to the cluster shapes, as sketched in Figure 2.7. Let us go through them in detail. Figure 2.7c shows results for the prolate (ellipsoid with one long and two short principal axes) cluster Na_{15}^+. Because the z axis (denoting the symmetry axis) is longer, the restoring Coulomb force is softer and the associated oscillation frequency (Mie plasmon along z axis) lies at lower energy, around 2.27 eV from the experimental data. In contrast, the two other directions orthogonal to the symmetry axis have shorter extensions thus associated with higher oscillation frequencies. Because there are two short axes and only one long axis (prolate shape), the higher-energy signal is twice as intense as the low-energy branch, as is clearly visible from the experimental distribution. The same conclusions hold for the other two clusters. In the case of the oblate Na_{19}^+ (2 longer axes and one smaller axis), the mechanism is the same, delivering this time one high-energy peak about 2.9 eV and a twice as intense (double) low-energy peak. Finally, Figure 2.7a displays the case of the (electronic shell closed) Na_{21}^+. The spherical nature of the cluster yields the same response frequency in each direction and so, forms one unique peak, characteristic of a spherical object. The considerations are

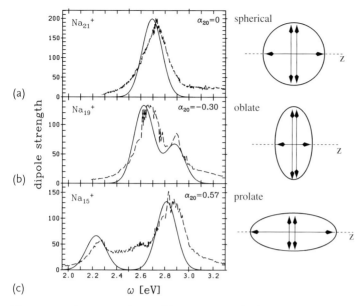

Figure 2.7 Deformation splitting of the Mie resonance in Na_{21}^+ (spherical, a), Na_{19}^+ (oblate, b), and Na_{15}^+ (prolate, c). The panels show optical absorption strengths for the three shapes, experimental from [34] and theoretical from linearized TDLDA on soft jellium background [35]. Sketches of the cluster shapes are also shown (deformations are exaggerated to emphasize the point). The lines with arrows indicate the directions of the basic plasmon oscillations. The symmetry axis (z axis) is indicated by a fine dashed line.

complemented by theoretical calculations (see Section 3.4.4) in which the deformations have been properly taken into account. The remarkable agreement between experimental and theoretical results further supports the interpretation of results as delivering a direct map of the cluster deformation. This analysis works reliably well for small clusters up to $N \approx 40$. For larger clusters, the optical response becomes increasingly blurred by coupling the Mie plasmon to individual dipole transitions (Landau fragmentation, see discussion of Figure 4.7). This inhibits a simple analysis of the underlying cluster shape.

A good understanding of the optical response of a cluster is the basis for studying further observables because the coupling to a laser field is the entrance door to nearly all dynamical scenarios. It is important to know where photoabsorption is large and where it is small. If the laser frequency matches an eigenfrequency of the cluster (as identified from the optical response), the coupling to light becomes resonant leading, for example, to an enhanced ionization. This reasoning holds for laser fields below the immediate breakup threshold. For even larger laser intensities, this frequency-sensitive (or photon-dominated) regime is progressively replaced by a field-dominated regime where field strength becomes more important than frequency. We shall discuss these strong field scenarios in more detail in Section 4.4, but show a first example in the next paragraph.

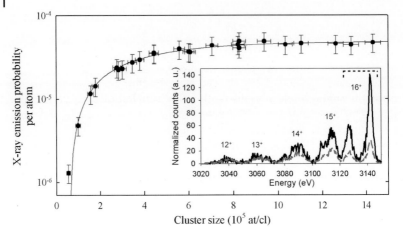

Figure 2.8 Evolution of the X-ray emission probability with Ar cluster size. The laser peak has an intensity of 3.4×10^{16} W/cm^2 and a pulse length of 70 fs. The insert provides charge state distributions for two cluster sizes in the saturation regime: dashed gray line for 6×10^5, and solid black line for 1.4×10^6. After [36].

A remarkable feature of the irradiation of clusters by intense laser beams is the surprising capability of clusters to absorb energy from the laser field, much more than what would occur in an atomic vapor. The obvious consequence of the high-energy accumulation is the emission of very energetic particles such as ions, electrons and also photons. The latter may come from atomic transitions, even involving deeply lying atomic levels, in the keV range, thus leading to X-ray emission. The case is illustrated in Figure 2.8 for irradiated Ar clusters. The high-energy absorption leads to high levels of ionization, not only for the cluster as a whole but also at the level of atomic constituents, reaching in that case (atomic) charge states of order 15 (not shown). The inset of Figure 2.8 shows the X-ray spectrum with high resolution in a narrow window of energies for K shell transitions (1s $\leftrightarrow$ 2p). The larger the X-ray energy the higher the charge state. One can identify several well-separated peaks, each one associated to a different charge state as indicated. Such a spectrum indicates a mix of highly charged ions in the excited cluster. The very intense laser pulse, amplified by the response of the valence electrons, has indeed efficiently removed many electrons from the Ar core states. Most of the thus released electrons are still kept in the cluster as a whole and float around there. This goes on for a possibly long time interval before complete explosion of the system due to the Coulomb pressure from the high net charge acquired by the cluster. Figure 2.8 furthermore shows the evolution of X-ray emission probability with cluster size. The average cluster size is determined by the backing pressure of the source (not indicated here), which can be related to an actual size (see Section 2.1.2). The emission probability clearly exhibits a saturation behavior. This is one more example of size dependencies of cluster properties (see discussion of Figure 1.7).

2.3.4
Electron Signals

The last class of measured particles we want to discuss are electrons, which provide detailed information on cluster structure and dynamics. Electron signals can be recorded at various levels of sophistication. The simplest case just records the total number of emitted electrons. This naturally yields an energy, angle, and time-integrated signal. This measurement is the complement of the ionic signal recording cluster charge states, at least in the cases of moderate ionization in which the cluster is not destroyed. This simple-most signal of net ionization becomes particularly useful in connection with pump-probe experiments (see, for example, Figure 5.6). The system is excited by a pump laser pulse. This often involves a slight ionization which, in turns, triggers ionic oscillations of the cluster by the changed Coulomb equilibrium. A probe laser pulse then explores the resonance conditions as a function of time delay where the enhanced ionization signal is exploited as an indicator of a resonance. Time-resolved measurements are, in principle, also possible with the more elaborate electron signals in PES and PAD. But this combines two demanding experimental setups and is thus more difficult to handle. Nonetheless, there exist some measurements on anionic Hg clusters and water clusters.

The next step is to measure the properties of the emitted electrons in detail. This can concern the kinetic energy or the direction of emission. Recording the kinetic-energy spectra of emitted electrons after irradiation by a laser with energy $\hbar\omega_{las}$ is called photoelectron spectroscopy (PES). Figure 2.9 illustrates the principle. The PES gives access to the single-electron energy ε_i of the initially occupied electron states i inside the cluster through the simple relation

$$\varepsilon_{kin} = \varepsilon_i + \nu\hbar\omega_{las} , \tag{2.7}$$

where ν is the number of photons involved in the process. The schematic view of PES is complemented in Figure 2.9c with an experimental example from Na_{58}^-. Such an anionic cluster has a rather low IP. All further valence electron states are also not strongly bound. Thus they can easily be turned to continuum electrons according to Eq. (2.7) with one photon of easily available frequency (in the visible). Such one-photon measurements on anionic clusters were thus already performed in the early 1990s, see for example the measurement of [11] reported in Figure 1.6. The steady improvement of light sources meanwhile allows to deliver high-energy photons and to also perform one-photon PES for neutral or cationic clusters, as those from neutral fullerenes and from positively charged metal clusters. Multi-photon ionization (MPI), as already indicated in Eq. (2.7) with $\nu > 1$, is possible as well, the laser pulses being highly coherent. For moderate laser intensities, this process maps in the PES further copies of the occupied electron spectrum with increasing kinetic energy, each copy separated by $\hbar\omega_{las}$. Further aspects of MPI will be discussed in Section 4.3.2.

The standard PES technique can also be improved by performing the so-called ZEKE (zero electron kinetic energy) measurements, as applied for example to neutral clusters. The idea in ZEKE is to tune laser frequencies so that electrons are

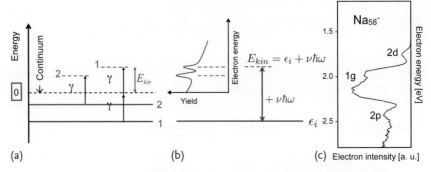

Figure 2.9 (a) Schematic view of photoelectron spectroscopy. The sample system has two single-electron states, $\epsilon_1 < \epsilon_2 < 0$. The IP is taken here as the energy of the HOMO, that is IP $= -\epsilon_2$. The emission threshold is taken as the reference of zero energy. The measured kinetic energies of emitted electrons are positive and recorded from threshold on upwards. The number of photons ν necessary for electronic emission does depend on photon frequency. In the case of the scheme, one photon suffices to emit from IP level but two are required to deplete the deeper lying state. (b) Each initially occupied state produces its own peak in the kinetic energy distribution, as indicated schematically. (c) An example is shown of a PES measurement for the Na_{58}^- cluster [37].

stripped with almost vanishing kinetic energy. To get rid of electrons possibly extracted with too large kinetic energies, one waits a while (typically 1 µs) so that fast electrons have had enough time to leave the system. This leaves only electrons with vanishing kinetic energy in the vicinity of the cluster. A low voltage pulse then accelerates these low-energy electrons and extracts them from the system. One scans laser frequencies and draws the yield as a function of them. This ZEKE technique provides a much higher resolution in photoelectron spectra (~ 0.1 meV) than standard photoelectron spectroscopy (~ 20–50 meV), but requires a much more involved setup. It is thus only marginally used.

Several techniques allow to measure the kinetic energies of the emitted electrons. Again the TOF provides here a versatile tool, but this time applied to the electrons themselves. As the mass to charge ratio is well defined for an electron, the arrival delay directly maps their kinetic energy. The much larger charge to mass ratio nevertheless requires careful guiding of the electron flow, for example, by a magnetic mirror [17]. Another interesting technique is provided by photoimaging spectroscopy, also known as the velocity map imaging (VMI) technique. This method uses a static electrical field to map the distribution of electron velocities onto definite positions on a detection screen, for a review, see [38]. This approach is extremely interesting as it allows a simultaneous determination of PES together with the photoelectron angular distribution (PAD). It is obvious that the measurement of angular distribution also brings valuable information on the cluster response (see Section 4.3.3). An example of a combined PES/PAD of electrons from irradiated Na_{58}^- is shown in Figure 2.10, showing a grayscale map of emission yield in the plane of kinetic energy and emission angle θ. The latter is measured with respect to the laser polarization. The figure demonstrates a clear dependence of the

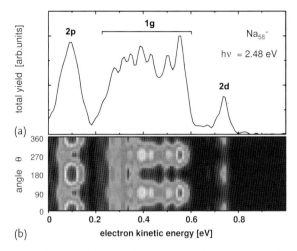

Figure 2.10 (a) PES/PAD of Na_{58}^- obtained with a photon energy of 2.48 eV. Standard PES (integrated over θ). The peaks are denoted by the corresponding single-electron states. (b) Combined PES/PAD obtained as a grayscale plot in the plane of electron kinetic energy and emission angle θ. Black stands for low emission intensity and white for maximum intensity. Adapted from [39].

photoemission on electronic wave functions. The mere PES in Figure 2.10a shows well-resolved peaks which can be associated by Eq. (2.7) to the single-electron states from which they are emitted by a one-photon process ($\nu = 1$), as indicated in standard spectroscopic notation (see Section 1.1.3.1). Note that the degeneracy of the 1g state is split into a series of subpeaks due to symmetry breaking of the ionic configuration. Figure 2.10b shows the combined PES/PAD. The comparison with the PAD shows that the 2p and 2d electrons are emitted parallel to the laser polarization, while emission from all the 1g states is preferentially aligned perpendicular to it. These remarkable results demonstrate that PAD add further useful information about the spatial structure of the emitting states (for more examples and discussion see 4.3.3). One should note that free clusters constitute an ensemble in which the cluster orientation with respect to laser polarization is unknown. What is measured is an average over all possible cluster orientations. From a theoretical point of view, this orientation issue has to be accounted for by appropriate orientation averaging.

3
How to Describe Clusters

A cluster represents a complex many-body system consisting of N atoms with their nuclei and all electrons. The exact solution of the Schrödinger equation for all electrons and nuclei is an extremely demanding task. For example, in a still small cluster as Ag_8, one has 96 constituents, 8 ions and 8×11 electrons, running at extremely different scales of time (see Section 1.1.2), length, and energy. Handling that complexity is beyond present day computing facilities, except perhaps for extremely small systems. Practical calculations have thus to employ more or less dramatic approximations. This is, in fact, the general problem of many-body quantum mechanics addressed long before in atomic, molecular and solid-state physics. Cluster physics is only one particular application, however a rather demanding one as we often deal with large systems without any symmetries. Thus one looks for particularly efficient methods and is easily ready for compromises to render calculations at all feasible. This chapter aims at providing a brief overview of theoretical approaches typically used for the description of free clusters. It first addresses the proper treatment of the cluster's components, namely electrons and ions, and the interactions between them. We finally discuss in a last section how typical observables can be computed, in order to be able to perform direct comparisons to experiments. An even more demanding problem is posed by clusters in contact with a substrate which calls for hierarchical approaches. This will be discussed in Section 5.2.

Before stepping down to the approximations, let us briefly recall the full many-body problem. The starting point is to choose the relevant degrees of freedom and their level of theoretical treatment. In Section 1.1.2, we explicitly distinguished (valence) electrons and ions (i.e., nucleus plus core electrons) in terms of time scales. This originates from the large mass difference between ions and electrons. Following this line, we separate the N_{el} electronic variables (as for example $\mathbf{r}_i$ for position of electron i) from the N_{ion} ionic ones, denoted by capital letters (as for example $\mathbf{R}_I$ for the ionic position and I for labeling of ions). We can now write the full Hamiltonian of the system as

$$\hat{H} = \hat{H}_{el} + \hat{H}_{coupl} + \hat{H}_{ion}, \tag{3.1a}$$

$$\hat{H}_{el} = \sum_i \frac{\hat{p}_i^2}{2m_{el}} + \frac{1}{2}\sum_{i \neq j}\frac{e^2}{|\hat{\mathbf{r}}_i - \hat{\mathbf{r}}_j|} + \hat{U}_{ext,el} = \hat{T} + \hat{W}_{ee} + \hat{U}_{ext,el}, \tag{3.1b}$$

An Introduction to Cluster Science, First Edition. Phuong Mai Dinh, Paul-Gerhard Reinhard, and Eric Suraud.
© 2014 WILEY-VCH Verlag GmbH & Co. KGaA. Published 2014 by WILEY-VCH Verlag GmbH & Co. KGaA.

$$\hat{H}_{\text{coupl}} = -\langle \Psi | \sum_{i,I} \frac{Z_I e^2 r^2}{|\hat{r}_i - \hat{R}_I|} | \Psi \rangle , \qquad (3.1c)$$

$$\hat{H}_{\text{ion}} = \sum_{I=} \frac{\hat{P}_I^2}{2M_I} + \frac{1}{2} \sum_{I \neq J} \frac{Z_I Z_J e^2}{|\hat{R}_I - \hat{R}_J|} + \hat{U}_{\text{ext,ion}} , \qquad (3.1d)$$

where $\hat{r}_i$ and $\hat{p}_i$ represent electronic position and momentum operators, and $\hat{R}_I$ and $\hat{P}_I$ the same for ions. $\hat{U}_{\text{ext,el}}$ and $\hat{U}_{\text{ext,ion}}$ stand for possible external potentials, for example, from a laser. An exact solution would solve the Schrödinger equation with this Hamiltonian for the full wave function $|\Psi\rangle \equiv \Psi(r, R, t)$, where we use here a compact notation R for the set of ionic coordinates and r for the electronic ones.

The Hamiltonian (3.1a)–(3.1d), although very involved, is already a simplification as it ignores electronic relativistic dynamics, magnetic effects and the spin-orbit force. Note that the latter plays a role in magnetic materials, as for example Cr or Fe clusters, and relativistic effects are important in heavy elements. Cluster magnetism represents a rich field with its own interest and is often explicitly studied. Nonetheless, we continue the presentation of theoretical approaches with the purely coulombic Hamiltonian (3.1a)–(3.1d) for reasons of simplicity. Note that the operator hat is usually omitted for the position r_i because we tacitly assume that we work in the local representation in most cases. The same holds for any local operator as, for example, a local potential $U(r)$.

3.1
Approximations for the Ions

3.1.1
The Adiabatic, or Born–Oppenheimer, Approximation

A fully quantum mechanical treatment of ions and electrons at the same level is extremely cumbersome and rarely done. One takes advantage of the fact that ions are much heavier than electrons, for example, $M_I/m_{\text{el}} \simeq 2000$ for H, 46 000 for Na or 420 000 for Pb. This allows one to treat the ions as classical particles which is a very helpful simplification. There exist basically two different routes of approximation: the simultaneous propagation of ions (classically) and electrons (quantum mechanically) or the adiabatic (Born–Oppenheimer) approximation. The first strategy will be addressed in Sections 3.1.2 and 3.1.3, while we now present the second one.

The Born–Oppenheimer (BO) approximation is a widely used standard in molecular physics and quantum chemistry [40]. The mass difference between ions and electrons causes a large difference in time scales (see Figure 1.3), so that electrons can follow almost instantaneously any change in ionic configuration. This allows one to decouple electronic motion from ionic motion, provided that electrons start and stay close to (instantaneous) equilibrium. This is the essence of the adiabatic approximation where the fully coupled many-electron–ion wave function Ψ is approximated by the adiabatic separation

$$\Psi(r, R) = \phi(R)\, \psi_R(r) . \tag{3.2}$$

$\phi(R)$ stands here for the ionic wave function. The electronic state $\psi_R(r)$ still depends parametrically on the actual ionic positions, hence "R" is written as an index. This allows one to decouple the full Schrödinger equation into two equations, one for ions and one for electrons (for details, see [40]). The scheme now proceeds in three steps. First, the electronic state $\psi_R(r)$ is determined for frozen ionic positions R as a stationary solution of the electronic Schrödinger equation:

$$(\hat{H}_{\mathrm{el}} + \hat{H}_{\mathrm{coupl}})|\psi_R\rangle = E_{\mathrm{el}}|\psi_R\rangle . \tag{3.3}$$

Second, we use this electronic wave function to compute an effective ionic potential V_{BO}:

$$V_{\mathrm{BO}}(R) = V_{\mathrm{ion}}(R) + \langle\phi(R)|\hat{H}_{\mathrm{el}} + \hat{H}_{\mathrm{coupl}}|\phi(R)\rangle . \tag{3.4}$$

Third, we solve the ionic Schödinger equation (stationary or time-dependent) using this V_{BO}.

It is interesting to note that the BO separation still allows for dealing with quantum mechanical ions. This is gratifying because quantum effects in molecular ionic motion can be crucial. They are, for example, required to obtain the observed discrete vibrational and rotational spectra [41]. The dynamics of vibrational wave packets is actually explicitly explored in pump-probe experiments [42]. One has even produced quantum mechanical interference patterns for scattering of rather large molecules [43]. All these processes deal with slow ionic motion, very well handled with the BO (or adiabatic) approximation.

3.1.2
Born–Oppenheimer Dynamics

3.1.2.1 Basic Equations of BO Dynamics

A quantum mechanical treatment of ions is not necessary in many dynamical scenarios. Ionic motion can then be classically treated. The assumption may be a bit critical for the lightest species of hydrogen molecules. But most practical calculations of fully dynamical processes in clusters up to now use classical ions. The ionic propagation is determined by the following classical equations of motion for the ionic positions R_I and momenta P_I:

$$\partial_t P_I = -\nabla_{R_I}\left[V_{\mathrm{ion}}(R_I) + \langle\psi_R|\hat{H}_{\mathrm{coupl}}|\psi_R\rangle\right], \quad \partial_t R_I = \frac{P_I}{M_I} . \tag{3.5}$$

The electronic wave function ψ_R is the stationary solution of the Schrödinger equation (3.3). It thus represents the electronic ground state for a given ionic configuration R. In principle, this electronic minimization has to be performed at each ionic time step. This procedure represents the BO molecular dynamics (BO-MD) where ions move on a BO surface (or potential energy surface), while electrons are thus evolving along the corresponding ground states.

3.1.2.2 Car–Parrinello Molecular Dynamics

Simultaneous propagation of ions and electrons can also be used as an efficient and conceptually simple realization of BO-MD. This is called the Car–Parrinello molecular dynamics (CP-MD) [44]. Instead of explicitly minimizing the stationary Schrödinger equation of the electrons, as done in BO-MD, electrons are propagated dynamically, but with an electronic pseudomass μ (typically several hundreds of the physical electron mass). The fictitious dynamics ensure that electrons quickly relax to the electronic ground state corresponding to each ionic configuration along the simultaneous ionic dynamics. This procedure is usually faster than the minimization of the stationary electronic Schrödinger equation (3.3) at every ionic time step. Note, however, that the pseudomass μ should not be too large to avoid spurious energy transfer between ionic and electronic degrees of freedom. One has to find a good compromise. Too small μ slows down the method. But with too large μ, adiabaticity is not guaranteed, and the full system does not follow a BO surface anymore. Note that CP-MD is sometimes denoted as *ab initio* molecular dynamics. The CP-MD has found widespread applications in cluster science, first for structure optimization (see Section 3.1.4) but also in adiabatic situations where the excitation remains small [45].

3.1.3
Beyond the Born–Oppenheimer Approximation

There are situations where the adiabatic approximation ceases to be applicable or becomes inefficient. This then calls for methods where both electrons and ions are propagated dynamically. At first glance, one may expect more complications. However, high excitations make a treatment of ions as classical particles even more legitimate. We will mention here two of the various methods developed thus far to deal with simultaneous propagation of (classical) ions and (quantum mechanical) electrons. It should be emphasized that there are still open problems. The case of dynamics beyond the adiabatic regime remains one of the major theoretical challenges in the field.

3.1.3.1 Simultaneous Propagation of Ions and Electrons

The BO approximation is by construction restricted to low-energy phenomena. In the case of large electronic excitations, such as delivered by intense lasers or by collisions with energetic ions, one has to explicitly account for electronic dynamics, allowing electrons to explore possibly highly excited electronic levels, that is far off the ground state BO surface. A typical example is a situation leading to ionization after laser irradiation (see Section 4.4). One of the most efficient approaches to deal with that is based on a time-dependent density functional theory (TDDFT) for electrons (Section 3.2.4), nonadiabatically coupled with MD for ions, for details see [46, 47]. The TDDFT-MD method applies to all dynamical situations. This includes the adiabatic limit and extends to cases which are far from the adiabatic limit, thus embracing truly diabatic scenarios. The practical realization combines the (simple) classical propagation to the quantum mechanical time evolution of the electronic states. Although technically straightforward, the actual propagation with TDDFT-MD can become cumbersome due to the dramatic span of time scales: the full resolution of electronic dynamics requires a very small time step (typically a fraction of 1 fs), whereas it takes several ps to explore global ionic dynamics (see Figure 1.3).

3.1.3.2 Trajectory Surface Hopping

The original BO-MD becomes invalid in the vicinity of level crossings, that is if two BO surfaces come close to each other during dynamics. The probability of transitions between electronic states from different BO surfaces can therefore become nonnegligible. The same holds for TDDFT-MD if electronic configurations of the same energy show up during dynamics. To cope with this situation, the BO-MD can be complemented by a "trajectory surface hopping" algorithm between energetically close electronic states [48]. This technique generates an ensemble of BO trajectories from which one can deduce reaction rates, averages and variances. It has been applied so far especially in "simple" systems with a well-controlled and limited number of competing BO surfaces and, accordingly, few possible hopping channels. An extension to TDDFT-MD is still a development task.

3.1.4
Structure Optimization

Before starting any dynamics, one has first to establish the electronic and ionic ground-state configurations. The stationary solutions for electrons are found by standard techniques [47]. The simultaneous optimization of the ionic configuration is more tedious because there are often many isomers around which have to be carefully discriminated (see for example Section 1.2.1). We briefly address the computation of the ground-state configuration in this section.

3.1.4.1 Minimum Energy Configurations by Dynamical Methods

A conceptually simple and straightforward optimization scheme can be deduced from a simultaneous time propagation of electrons and ions according to TDDFT-MD as described in the previous section. The aim here is not to simulate a fully coupled electron-ion dynamics but to cool the combined ion-electron system down into its ground-state configuration. This is achieved by removing once in a while kinetic energy from the ions by simply setting $P_I \to 0$. This steadily extracts energy until the system comes to an halt in a minimum energy configuration. To find appropriate times for resetting P_I, one can take a protocol of the ionic kinetic energy $E_{\text{kin,ion}}$ and best resets the ionic momenta if $E_{\text{kin,ion}}$ has reached its maximum. This means that a maximum amount of energy is extracted with one cooling step.

However, the huge mass difference between ions and electrons makes this method rather slow because many electronic cycles are computed until one explores a significant change of ionic motion. One can exploit the fact that the electron cloud readjusts extremely quickly to the instantaneous ionic configuration and thus we can reduce the mass ratio by one or two orders of magnitude. This is legitimate because we are not here interested in the dynamical features but only in the end point of the relaxation which is independent of the mass ratio. This leads eventually to a Car–Parrinello type propagation (see previous section) here combined with cooling for structure optimization.

3.1.4.2 Minimum Energy Configurations by Stochastic Methods

All direct cooling methods, as those explained in the above paragraph, have the problem that they can easily get stacked in a local minimum, that is in an isomeric configuration. To map the variety of isomers and to find the true ground state amongst them, one has to restart the cooling from a large variety of initial configurations.

A more systematic way to encircle the minimum and/or to map the whole landscape of isomers is provided by simulated annealing with Monte Carlo (MC) techniques [49], widely used in all areas of physics. It consists of a random walk through the possible stationary states of the system. In our case, a state is characterized by a point R in the $3N_{ion}$-dimensional coordinate space of the ions together with the wave function ψ_R of the N_{el} electrons. The key quantity for the walk is the total energy $E(R, \psi_R(r))$ which is a functional of the state. Only this energy is needed, no forces and no explicit functional derivatives. Random changes of the state are checked for their change in energy. A lowering is always accepted. A step with increasing energy is occasionally followed to avoid being trapped in a secondary minimum.

3.1.5
Approaches Eliminating Electrons

3.1.5.1 From Ions and Electrons to Atoms

In many situations, a detailed account of electrons *and* ions separately is not compulsory. This applies to cases in which one is interested only in changes at the length scale of ionic positions and not so much in the finer details of the electronic distributions. For then, only the electronic contribution to global binding and the electronic transport between ionic sites do matter, and this suggests a minimalistic account of electronic degrees of freedom. This idea leads to a wide class of models in which the effect of electrons is directly plugged into a potential acting between ions/atoms. The archetypes are Hückel or tight binding (TB) models [41]. The electronic wave functions are here treated at the level of linear combination of atomic orbitals (LCAO) using a small basis of atomic orbitals. Thus the overall computation is rather simple, and yet highly accurate. This allows for large-scale/long-time Born–Oppenheimer computations, which are otherwise, hard to envision.

The TB method is especially valuable for covalent binding where electrons remain rather localized. It has also been refined to become applicable to metallic clusters. For then, polarization effects are added to the picture to produce the distance-dependent tight binding (DDTB) approach [50]. Note also that the TB method has a strong relationship with the much celebrated Hubbard model, widely used for describing electrons in solids [51, 52] and quantum liquids [53]. Again, the Hubbard model is well suited to deal with localized electrons and once complemented by polarization effects, it has proven to be a versatile tool for cluster structure problems as well [54].

3.1.5.2 Fully Classical Molecular Dynamics Approaches

There are many situations where one can even simplify the TB approach by reducing the number of atomic orbits to one per atom. This provides a particularly simple picture, reducing the problem to the mere treatment of atoms taken as classical point particles interacting by effective atom–atom potentials. Such models are well established in molecular physics and constitute the class of classical molecular dynamics (MD) approaches. The prototype class of systems easily attainable by such methods are rare gas clusters where the effective potential can be very well parametrized in the Lennard-Jones form [10]

$$V_{\text{Ar-Ar}}(r) = V_0 \left[\left(\frac{\sigma_{\text{vdW}}}{r} \right)^{12} - \left(\frac{\sigma_{\text{vdW}}}{r} \right)^{6} \right], \quad (3.6)$$

where the parameters are, for the case of Ar, $V_0 = 4 \times 7.6 \times 10^{-4}$ Ry $= 480$ K, $\sigma_{\text{vdW}} = 6.43\, a_0$ and $m_{\text{Ar}} = 7.35 \times 10^5\, m_{\text{el}}$. What remains is simple classical dynamics for the atoms following Eq. (3.5), but now of course without the ion–electron coupling. The MD approaches allow large-scale studies of cluster dynamics as, for example, the computation of thermodynamic properties and phase transition [55].

The case of van der Waals clusters with simple atom–atom potentials has been extensively studied for many years, in particular for systematic studies of temperature effects in clusters. But MD potentials have been developed to a high degree of sophistication, even with polarization effects and allowing nowadays studies of various combinations of materials including metals [56]. Figure 3.1 illustrates the versatility of MD in the case of impact of Ar clusters on a Ni surface. The simplicity of the MD approach allows in that case to simulate rather large clusters in contact with a surface which is modeled by a sizable amount of atoms. Note also the long simulation times of several picoseconds.

3.2
Approximation Chain for Electrons

Many-electron systems are highly correlated, and exact calculations of their properties are extremely involved, mostly beyond feasibility. Therefore, a broad selection of approaches has been developed, each one being a compromise between precision and expense. Different situations require different compromises and thus we

3 How to Describe Clusters

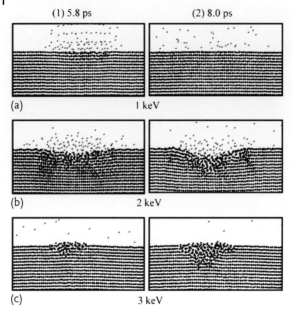

Figure 3.1 Example of molecular dynamics simulation: snapshots of Ar$_{353}$ cluster impacts on a Ni(111) surface with different incident energies of 1 keV (a), 2 keV (b) and 3 keV (c) at 0.8 ps (1) and 3.0 ps (2) after impact. From [57].

find a broad palette of methods in use. In this section, we present the most widely used schemes, paying particular attention to modeling based on density functional theory (DFT) which seems to be presently one of the most efficient tools in cluster dynamics.

3.2.1
Overview of Approaches for the Electronic Subsystem

Figure 3.2 presents the most widely used theoretical approaches and sketches the regimes of their applicability in the plane of excitation energy and particle number. Since the decision for a method depends on several other aspects (e.g., demand on precision, material, time span of simulation), the boundaries of the regimes are to be understood as very soft with large zones of overlap between the models.

The most elaborate models are the *ab initio* methods which start from the basic Hamiltonian (3.1a)–(3.1d) and derive everything from there, within a given choice of trial wave functions. A widely used scheme is here the configuration interaction (CI) method which relies on an expansion into a superposition of Slater states, see Section 3.2.2 and (3.8). The limitations for CI (and other *ab initio* methods) are purely a matter of practicability. The range of applicability will slowly grow with the steadily increasing computer power.

Density functional theory (DFT) is a mean-field theory, similar to Hartree–Fock (see Section 3.2.2), which describes a system effectively in terms of a set of single-

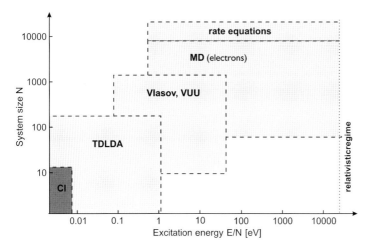

Figure 3.2 Schematic view of applicability of different approaches (see text for details) in a landscape of system size versus excitation energy per atom.

electron occupied states (see Section 3.2.4). It is mostly realized as the local density approximation (LDA) or time-dependent LDA (TDLDA). These approaches are limited in system size for practical reasons and in excitation energy for physical reasons, because of the missing dynamical correlations from electron–electron collisions (developing at intermediate time scales, see Figure 1.3).

A semiclassical mean-field description is provided by the Vlasov equation originally designed for plasma physics [58]. One can define the Vlasov mean-field with LDA and thus obtain Vlasov-LDA as the semiclassical limit of TDLDA. This limit will be detailed a bit in Section 3.2.5. This ignores quantum effects as shell structure or tunneling and thus limits applicability at low energies. On the other hand, the semiclassical treatment allows to include dynamical correlations due to electron–electron collisions leading to the Vlasov–Uehling–Uhlenbeck (VUU) approach which extends the applicability to larger energies than those allowed by TDLDA.

Even higher excitations and system sizes are the realm of electronic molecular dynamics approaches and rate equations which, however, are even more limited than VUU for low energies and small systems. The upper limit in energy is given by the onset of the relativistic regime, where retardation effects within the coupling begin to severely influence the dynamics.

3.2.2
Ab initio Methods

There is a large class of approaches denoted as *ab initio*. In a broad sense, it embraces all techniques which can be derived more or less strictly from the initial Hamiltonian (3.1a)–(3.1d). For example, density functional methods discussed in Section 3.2.4 are often called *ab initio* although the relation to (3.1a)–(3.1d) is sophisticated. In a narrower sense, methods are *ab initio* if they directly employ the Hamiltonian (3.1a)–(3.1d) and deal with it under well-defined approximations, in particular in the choice of the Hilbert space for the electronic wave functions.

3.2.2.1 Hartree–Fock (HF)

The lowest level *ab initio* method is the Hartree–Fock (HF) approach [28]. It is derived by making an ansatz for the N-body wave functions as an antisymmetrized product of independent single-electron states (Slater determinant):

$$\Psi(r_1,\ldots,r_N) \approx \Phi_0(r_1,\ldots,r_N) = \mathcal{A}\{\varphi_1(r_1)\ldots\varphi_N(r_N)\}, \quad (3.7)$$

where $\mathcal{A}$ is the N-body antisymmetrization operator. Optimization of the single-particle (s.p.) wave functions φ_i is performed by applying the Ritz variational principle to the expectation value of the Hamiltonian (3.1a)–(3.1d) [41]. The emerging self-consistent HF equations look like a one-body Schrödinger equation, $\hat{h}_{\mathrm{HF}}\varphi_i = \varepsilon_i\varphi_i$ for each occupied s.p. state φ_i. However, the effective HF mean-field Hamiltonian $\hat{h}_{\mathrm{HF}}$ depends on the set of s.p. states, which makes the HF equations highly nonlinear. The generalization to dynamical processes is time-dependent HF (TD-HF) which can be derived from the ansatz (3.7) by a time-dependent variational principle [59], leading to the TDHF equations $\hat{h}_{\mathrm{HF}}\varphi_i = i\hbar\partial_t\varphi_i$ where $\hat{h}_{\mathrm{HF}}$ has formally the same structure as in the stationary case. TDHF uniquely determines the time evolution $\{\varphi_i(r,t)\}$ for given initial conditions $\{\varphi_i(r,t=0)\}$. The (TD)HF equations are the most elaborate ones that one can do within an independent particle model. Their solution is already a demanding task due to the nonlocal exchange potential. There exist a few studies using HF in connection with clusters, see for example, [60, 61]. It is experience from atomic and molecular physics that HF is reliable and performs well. But correlations beyond mean field are considered to be important for a more satisfying agreement.

3.2.2.2 Configuration Interaction (CI)

The most widely used *ab initio* method beyond HF is configuration interaction (CI). It describes a correlated state Ψ through a sum of Slater determinants Φ_n as

$$|\Psi\rangle = \sum_n c_n |\Phi_n\rangle. \quad (3.8)$$

For a given set $\{\Phi_n\}$, the superposition coefficients c_n are variationally determined. This is a most general formulation of correlations which can be driven, in principle, up to an exact solution. Much depends on how one actually chooses the basis. For an extensive discussion on the many variants actually used, see, e.g., [62]. Standard CI uses a natural hierarchy of Φ_n starting from the HF state Φ_0 as given in (3.7), then adding one-particle-one-hole (1ph) states about Φ_0 up to a given energy, then 2ph states, and so on, until one finds sufficient convergence of the results. Such CI is widely used in this form in quantum chemistry and has also found several applications in cluster physics [63]. Still, the necessary size of the basis can grow huge and the CI method become very expensive. But the appeal of CI is that it

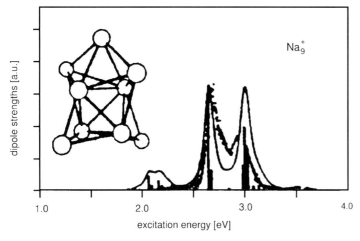

Figure 3.3 Photoabsorption strength for Na$_9^+$ computed with configuration interaction (CI). The insert indicates the structure of the cluster. The vertical bars show the results from the CI calculation. The full line is obtained by smoothing the discrete data. The dots indicate the experimental strength. After [64].

emerges as a systematic improvement beyond HF and that all compromises are concentrated in the well-controllable choice of the expansion basis. In fact, carrying the expansion far enough allows to approach an exact calculation and there exist many such large-scale CI calculations in atomic and molecular physics.

These elaborate CI calculations are also used to compute excitation spectra of clusters. In principle, the diagonalization of the CI Hamiltonian automatically produces excited states. In practice, an appropriate description of excitations requires an even more elaborate choice of the basis set. The necessary extensions depend very much on the actual system (symmetries, density of states). For large clusters, one may even decide to use approximations, as for example a linear response theory built on the CI ground state [63]. There are several applications of CI to compute optical response along these lines: an example for the optical response of Na$_9^+$ is shown in Figure 3.3. First, we note the good agreement with the experimental data. The detailed substructures dealt with in CI produce a great manifold of states (vertical bars). However, the basic pattern obtained after folding (solid line) reveal a rather simple structure of a dominant double peak around 3 eV and a smaller peak at 2 eV. This simple structure is nicely reproduced by TDLDA (not shown here). Remember that TDLDA is much easier to use than full CI and can be extended to the nonlinear domain, while applications of CI all remain necessarily in the linear domain of small amplitude motion. A truly time-dependent CI for large amplitude processes has, to our knowledge, not yet been tackled.

3.2.2.3 Alternatives to Configuration Interaction

A more refined strategy to optimize the basis states for the superposition (3.8) is multiconfigurational HF (MCHF), where the selection of all Φ_n together with the c_n is determined by variation. Such an optimized expansion converges much faster but leads to much more involved coupled equations. Thus MCHF is still in a developing stage. There are also developments for a time-dependent MCHF (MCTDHF) which is even more involved, but is being intensively studied at present [65].

Exact solutions of the many-electron Schrödinger equation are feasible for the homogeneous electron gas and for very small systems. A fully exact numerical solution for the homogeneous electron gas at zero temperature is found in [66]. The development has gone further to finite temperatures and nonhomogeneous systems. The first zero-temperature calculations from [66] are still the basic input for practically working LDA functionals (Section 3.2.4.4). At the side of molecules, there exist several exact calculations for structure, see for example rather recent calculations for carbon clusters up to fullerenes [67]. One can meanwhile even find fully dynamical exact calculations for very small systems, for example for the laser-induced dynamics of the He atom [68]. The progress of computer technology will steadily extend the range of applicability for exact calculations of many-electron systems. Yet, exact methods will be limited for a long time.

3.2.3
Phenomenological Electronic Shell Models

Early stages of such methods were concerned with a principle understanding of the shell structure and optical response of metal clusters. An educated guess for the mean-field potentials, called a shell model, was an extremely valuable tool for this. These phenomenological models are not well suited for cluster dynamics. Therefore, they are not much used anymore and do not appear in the scheme 3.2. Nonetheless, we briefly summarize here two of the models.

Considering bulk metal in the jellium approximation (see Section 3.3.2), we encounter a constant potential inside the material and a soft surface transition to zero potential. The same picture transferred to a spherical and neutral cluster suggests a Woods–Saxon profile:

$$U_{\mathrm{WS}}(\mathbf{r}) = -U_0 \left[1 + \exp\left(\frac{|\mathbf{r}| - R(\vartheta, \phi)}{\sigma_{\mathrm{WS}}} \right) \right]^{-1} , \qquad (3.9)$$

where U_0 is taken as the average binding potential in bulk matter. The Woods–Saxon shell model has been used in several early studies, in particular in metal clusters [69].

A substantial further simplification is achieved by realizing that shell effects are determined by the states near the Fermi surface and that these states practically see a harmonic potential. This suggests to use for first estimates a simple harmonic oscillator shell model [28]. The harmonic oscillator predicts spherical magic electrons shells at $N_{\mathrm{el}} = 2, 8, 20, 40, 70, 110, \ldots$ It fits observed shells in Na clusters up to $N_{\mathrm{el}} = 40$. Larger Na clusters (and most other alkalines) continue with 58, 92, and 138. The applicability can be extended to higher shells by down-shifting the states with high orbital angular momentum l using a phenomenologically fitted $\hat{l}^2$ term in the model, yielding the Clemenger–Nilsson model (see Section 4.1.2). Tuning the oscillator curvatures along the three principal axes separately allows to accommodate deformed situations including triaxiality, see [17] for an extensive review of early applications.

The computational simplifications of the Clemenger–Nilsson model are dramatic. The spherical model can be treated fully analytically. The deformed case requires only one simple diagonalization where at least all matrix elements are given analytically. The model remains useful to understand shell structure and deformation systematics in simple terms. However, it is not so suitable for dynamical studies. They also require the knowledge of the residual interaction between the single-electron states and their phenomenological parametrization would introduce too many new unknowns to be fixed.

3.2.4
Density Functional Theory

Density functional theory (DFT) methods have already been mentioned at several places, in particular in coupled ionic and electronic dynamics (see Section 3.1.3.1). The goal of DFT is to develop self-consistent equations which employ purely local effective potentials for the contributions from exchange and correlation. These potentials are to be expressed in terms of the local electron densities $\rho(r)$ of the system. DFT is thus more than HF in the sense that correlations are included. But it is less than HF since exchange is treated only approximately. In any case, it is much simpler to handle than HF, and even more so than CI, because mostly local potentials are dealt with and because independent single-electron states are involved. The success of the method depends on a diligent choice of these effective potentials. DFT is a huge field in itself. We take here a practitioner's approach and start the presentation of the Kohn–Sham (KS) scheme from a given energy functional. We do not address the theoretical foundations in terms of the much celebrated Hohenberg–Kohn theorem [70] and Kohn–Sham scheme [71]. We also skip many details of the derivation and implementation. That information is well documented in several books and reviews, for example in [72–74].

3.2.4.1 The Energy Functional for the Kohn–Sham Scheme
The Kohn–Sham (KS) scheme separates the total electronic energy into the kinetic term and the interacting one. One represents the N (valence) electrons, by N non-interacting Kohn–Sham (KS) orbitals (or s.p. states) $\varphi_i(r)$, $i \in \{1, \ldots, N\}$. The total electronic density is represented by these KS orbitals as

$$\rho(r) = \sum_{i=1}^{N} |\varphi_i(r)|^2 \ . \tag{3.10a}$$

The kinetic energy $E_\text{kin} = \langle \Psi | \hat{T} | \Psi \rangle$, where $\hat{T}$ is the kinetic energy operator, see (3.1b), is mapped to a sum of s.p. kinetic energies as

$$E_\text{kin}(\{\varphi_i\}) = -\frac{\hbar^2}{2m} \int d^3 r \sum_{i=1}^{N} \varphi_i^*(r) \nabla^2 \varphi_i(r) \ . \tag{3.10b}$$

This expression is a functional of the s.p. orbitals φ_i which serves to maintain the quantum shell structure in the KS calculations. The nontrivial correlation part of the exact kinetic energy is summarized in the interaction energy. The interacting term $\langle \Psi | \hat{W}_{ee} | \Psi \rangle$ uses the electron–electron operator $\hat{W}_{ee}$ defined in (3.1b). It is mapped to the density functionals $E_H[\rho] + E_{xc}[\rho]$. The first term E_H is the standard (direct) Coulomb Hartree energy, which is naturally a functional of ρ,

$$E_H[\rho] = \frac{e^2}{2} \iint d^3r \, d^3r' \, \frac{\rho(r)\rho(r')}{|r-r'|} = \frac{1}{2} \int d^3r \, \rho(r) \, U_H[\rho] \, . \quad (3.10c)$$

The second term E_{xc} is the exchange-correlation energy which accumulates all pieces of the exact energy not yet accounted for. It is the problematic part in the scheme, since its functional expression is not exactly known. Many approximations therefrom do exist, among which the simplest and most robust is the Local Density Approximation (LDA), see discussion below. We have furthermore the energy E_{coupl} from the coupling to the ions (see Section 3.3), and possibly the energy E_{ext} from an external electromagnetic field, which read as

$$E_{\text{coupl}} = \int d^3r \sum_{i=1}^{N} \varphi_i^*(r) \, \hat{V}_{\text{coupl}} \, \varphi_i(r) \, , \quad E_{\text{ext}} = \int d^3r \, \rho(r) \, U_{\text{ext}}(r) \, . \quad (3.10d)$$

Both these contributions couple to single electrons and are naturally well represented by an independent particle picture in terms of φ_i. All in all, we end with the following expression for the total electronic energy,

$$E_{\text{total,el}}[\rho] = E_{\text{kin}}(\{\varphi_i\}) + E_H[\rho] + E_{xc}[\rho] + E_{\text{coupl}} + E_{\text{ext}} \, . \quad (3.10e)$$

3.2.4.2 The Kohn–Sham Equations

The stationary KS equations are derived by variation of this energy with respect to the s.p. wave functions φ_i^*, yielding

$$\hat{h}_{KS}[\rho] \, \varphi_i(r) = \varepsilon_i \, \varphi_i(r), \quad (3.11a)$$

$$\hat{h}_{KS}[\rho] = -\frac{\hbar^2 \nabla^2}{2m} + U_H[\rho] + \hat{U}_{xc}[\rho] + \hat{V}_{\text{coupl}} + U_{\text{ext}} \, . \quad (3.11b)$$

The exchange-correlation potential is a standard functional derivative $\hat{U}_{xc} = \delta E_{xc}/\delta \rho$. Coupling potentials to ions and that to the external field are trivially given.

The time-dependent KS equations analogously read as

$$i\hbar \, \partial_t \varphi_i(r,t) = \hat{h}_{KS}[\rho] \varphi_i(r,t) \, , \quad (3.11c)$$

where $\hat{h}_{KS}$ is composed in the same manner as above, provided that one replaces $\rho(r)$ by $\rho(r,t)$. This assumes an instantaneous adjustment of the total electronic density, although memory effects can play in some cases an important role, especially in E_{xc} [73].

A few words are in order here concerning the external potential U_{ext}. It generally accounts for a time-dependent perturbation, as that delivered by a laser pulse or by a colliding projectile. It is however not required in stationary calculations or in simple excitation spectra (linear response). In turn, it may play a crucial role in the time-dependent KS equations when truly time-dependent processes are under study.

The stationary KS equations (3.11a) pose an eigenvalue problem. They provide the electronic ground state of a system. This is a highly nonlinear problem due to the self-consistent feedback of the local density in the KS Hamiltonian. It is usually solved by iterative techniques [46]. The time-dependent KS equations imply an initial value problem. The natural starting point is the ground state obtained from the stationary KS equations. The time-dependent KS system can then be solved by standard methods of first order differential equations [46]. Note finally that we wrote spinless KS equations. One can easily include the electron spin in Eqs. (3.11a)–(3.11c). We refer the reader to [72–74] for more details.

3.2.4.3 A Practical Example

One advantage of the KS scheme is to maintain a quantum shell picture in combination with a density functional approach. Figure 3.4 shows the case of Na_{58} as a typical example. The ionic background is handled by a soft jellium (see Section 3.3.2) and the exchange-correlation is treated by LDA. Figure 3.4a shows that jellium and electron densities indeed stay very close to each other. This minimizes the direct Coulomb energy. There remains an irreducible discrepancy in that the electron density has unavoidable fluctuations due to shell effects. One sees the

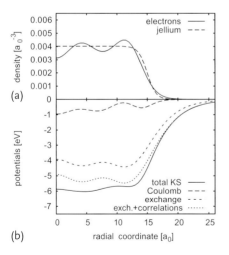
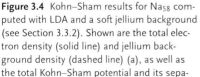
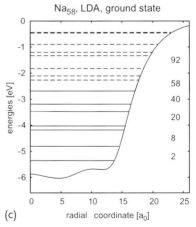

Figure 3.4 Kohn–Sham results for Na_{58} computed with LDA and a soft jellium background (see Section 3.3.2). Shown are the total electron density (solid line) and jellium background density (dashed line) (a), as well as the total Kohn–Sham potential and its separate contributions (b). Also illustrated is the sequence of single-particle levels (c). The solid lines denote occupied states, whereas dashed lines represent unoccupied ones. Magic numbers associated with shell gaps are indicated on the right side of the plot.

oscillatory pattern of the last filled shell. Its wavelength can be estimated from the Fermi momentum $k_F = (9/4\pi)^{1/3} r_s^{-1}$ [28]. The r_s therein is the Wigner–Seitz radius and $r_s \approx 3.9 a_0$ for Na [75]. This corresponds to a spatial oscillation frequency $\pi/k_F \approx 2\pi a_0$, as observed in the figure. Figure 3.4b shows the KS potential and its separate contributions. The part labeled "Coulomb" is actually the total Coulomb field from electrons and jellium background. It is extremely small as expected from the close matching of the densities. The Coulomb field may be a bit larger if a detailed ionic structure is considered. But even in this case, it remains a small part. The largest contribution to binding of neutral metal clusters comes from the Coulomb exchange energy and the correlation energy adds another 20–30% to the binding. Figure 3.4c shows the shell structure in terms of the sequence of single-electron levels. We see the large gap between occupied and unoccupied states which indicates that $N_{el} = 58$ is an electronic shell closure. We will work out in Section 6.1 that metal clusters are saturating systems, that is systems which develop along a nearly constant average density. Accordingly, the picture will look very similar for other cluster sizes (see Figure 6.2). Average density, potential depth, and the Fermi energy are insensitive to the particle number. However, more levels fit into the span between the lowest state and Fermi energy, which means that the level density increases with increasing system size.

3.2.4.4 The Local Density Approximation (LDA)

The general theorems of DFT guarantee that there exists a universal energy-density functional, but do not provide the way in which to determine that functional. Furthermore, it is likely that such a functional would have an extremely involved mathematical structure. To make DFT a practicable scheme, one needs to invoke approximations. We illustrate this point on the most widely used and very efficient Local Density Approximation (LDA) to the exchange-correlation energy. In fact, we have to deal with the local spin densities ρ_σ for the two spin orientations $\sigma \in \{\uparrow, \downarrow\}$. This is sometimes coined as the local spin-density approximation (LSDA). We use here the generic name LDA throughout and skip the spin index for the sake of simplicity. Note that there exist other elaborate approximations, often motivated by defects of the LDA. We will address this point at the end of this section.

The construction of LDA is simple. One computes the ground state of the homogeneous electron gas as exactly as possible. This yields the exchange-correlation energy per particle $E_{xc}/N \equiv \epsilon_{xc}$ as a function of the (yet homogeneous) density ρ. We rewrite that to the energy density (energy per volume), $E_{xc}/V = \rho \epsilon_{xc}$. The final and crucial step is to extend the validity of this global procedure to a local and time-resolved scale. This is done by allowing for an inhomogeneous and possibly time-dependent $\rho(\mathbf{r}, t)$ in that expression. It amounts to considering the energy as composed piecewise from an infinite electron gas of densities $\rho(\mathbf{r}, t)$. This is, at first glance, a bold approximation. It is valid at least for very gently varying densities, as it is typical for example for valence electrons in metals. In practice, however, LDA provides a good description for a wide variety of systems. There is an enormous body of literature pondering the successes and failures of this description, for a more detailed discussion, see for example [72]. Note that a functional depending on $\rho(\mathbf{r}, t)$ employs the instantaneous density and thus excludes any memory effect. This time-dependent generalization is often called adiabatic LDA (ALDA). Again, we use here the generic notation LDA.

As an example, let us consider the simpler case of the exchange functional in LDA and its contribution to the KS potential. The Coulomb exchange energy of a homogeneous electron gas has a simple analytical form $E_x = \int d^3r \rho(\mathbf{r}) \epsilon_x^{(\infty)}[\rho]$ with $\epsilon_x^{(\infty)}(\rho) \propto \rho^{1/3}$, which leads to an exchange potential of the form

$$U_x^{(\text{LDA})}[\rho] = \frac{\delta E_x[\rho]}{\delta \rho(\mathbf{r}, t)} = -e^2 \left(\frac{3}{\pi}\right)^{\frac{1}{3}} \rho^{\frac{1}{3}}(\mathbf{r}, t) \,. \tag{3.12}$$

This is the widely known Slater approximation to the exchange potential which was proposed and employed as a practical scheme long before the theoretical foundation of DFT.

Although LDA looks straightforward and unique, there are various (slightly) different exchange-correlation functionals available depending on different inputs from the electron gas. Fully exact numerical calculations have become available since the beginning of the 1980s. They have been taken as a basis for the widely used functional of [76]. Note that the differences between the LDA functionals mainly concern the correlation part. The exchange part in LDA is always the long established Slater approximation plus appropriate spin extension.

A word on nomenclature is in order: the application of LDA in the time domain is usually called time-dependent LDA (TDLDA). In general, TDLDA poses an initial value problem and can be driven at any rate of excitation. Early TDLDA calculations employed a linearization of the TDLDA equations and thus were restricted to small amplitude motion [77]. The linearized TDLDA which is often also called a random-phase approximation (RPA) [28].

3.2.4.5 DFT beyond LDA

The validity of LDA is a much discussed problem. It depends very much on the system under consideration, on the demands of precision aimed at, and on the studied observables. One of the remaining problems is the self-interaction error: the single-particle state φ_i is included in the density ρ, and thus contributes to the mean-field Hamiltonian which acts on φ_i. This yields wrong asymptotics for the Coulomb mean field which, in LDA, exponentially decays at large distances instead of behaving as $\propto e^2/r$ as it should. An attempt to reduce the problem was the generalized-gradient approximation (GGA) by augmenting LDA with an additional dependence of the exchange-correlation potential on $\nabla \rho$. GGA yields a significant improvement in the computation of atomic and molecular binding. For example, it lifts the description of dissociation energies to a quantitative level.

However, GGA does not fully remove the self-interaction error. To that end, a simple and robust self-interaction correction (SIC) was introduced in which the self-interactions are subtracted from the DFT energy. The self-interaction corrected KS equations are then derived again by variation. A problem is that the emerging KS Hamiltonian becomes state-dependent. The treatment of the SIC is generally cumbersome, in particular in a time-dependent KS scheme, for a review see [78] and for practical approximations [46, 47].

3.2.5
Semiclassical Approaches

Classical particles are easier to handle than quantum mechanical wave functions. This is already exploited when treating the cluster ions as classical particles, see the basic Hamiltonian (3.1a)–(3.1d). One also simplifies the treatment if one can describe the electrons classically as well. Such an approach will become increasingly valid with increasing excitation and/or temperature. The simplest picture emerges if one treats the electrons as classical point particles. That is done regularly in plasma physics and it has been shown there that the fully classical treatment is valid for temperatures $T \gg 4\text{Ry}(r_s/a_0)^{-2}$ [79]. That is a rather large temperature which rules out the majority of applications in cluster dynamics, except in extremely high-energy events following irradiation by intense laser beams (see Figure 3.2), where genuine plasma physics models are applied [80].

A larger range of validity applies if one considers a classical limit to a phase space distribution $f(r, p)$. This yields the widely known Vlasov equation [8] which is applicable, in principle, down to temperature and/or excitation zero. A further reduction fixes the momentum dependence in $f(r, p)$ to a Fermi-distribution profile. This then leads to the Thomas–Fermi approximation which is particularly suitable for stationary states, but may also be extended to moderately dynamical situations. Both approaches provide a pertinent description of the average cluster properties, but merely miss the quantum mechanical shell effects. The latter ones are known to disappear at around $T \sim \epsilon_F/20$ for metal clusters, for example $T \sim 1000–2000$ K for Na clusters. Above that point, the phase-space description becomes fully valid. At lower temperature, one nevertheless obtains at least a good picture of gross features [30]. The Vlasov equation is better suited to complex dynamical situations and it can be augmented by an Uehling–Uhlenbeck correlation term, thus incorporating dynamical correlations. The combination is called the Vlasov–Uehling–Uhlenbeck (VUU) scheme and it has been much used in the past decades for nuclear dynamics [81–83]. There exist several successful applications of Vlasov-LDA in the dynamics of simple metal clusters, see [84] and citations therein.

3.3
Approximation Chain for the Ion–Electron Coupling

As a last ingredient in (3.1a)–(3.1d) or (3.11a)–(3.11c), we now address the modeling of the coupling between quantum valence electrons and classical ions. Two different strategies are in order here: either we treat ions explicitly via pseudopotentials (see Section 3.3.1), or we smooth out the ionic structure and consider the ionic background in a jellium approach (see Section 3.3.2).

3.3.1
Pseudopotentials

3.3.1.1 Formal Foundation of Pseudopotentials (PsP)

Once having separated inert core electrons from reactive valence electrons, one wants to develop a scheme where only valence electrons are explicitly treated. This is achieved by pseudopotentials (PsP). We give here a brief summary of the idea behind PsP. A detailed introduction can be found in [85]. Let us assume that we have one isolated atom with N_c core electrons in single-electron states $\{\varphi_1, \ldots, \varphi_{N_c}\}$ and one valence electron in state φ_v. The states are supposed to be generated by some mean-field Hamiltonian $\hat{h}_{mf}$, such as given for example in Hartree–Fock or Kohn–Sham (see Section 3.2). We now augment the mean-field Hamiltonian by a PsP, $\hat{h}_{mf} \to \hat{h}_{mf} + \hat{V}_{PsP}$ such that core states do not appear again as solutions, and the new ground state ψ represents the valence state φ_v. This state ψ may then differ from the original valence state φ_v. In particular, it does not need to be orthonormal to the core states. But it is essential that ψ carries in the end the same physics in terms of energy and long-distance behavior. The core states accounted for in $\hat{V}_{PSP}$ provide a repulsion which repels ψ outside the core region, as would "naturally" occur for a true valence wave function, but this now occurs for a "pseudo" wave function, possibly without nodes. The $\hat{V}_{PSP}$ are in general nonlocal and energy-dependent operators. In practice, one takes these formal ideas to motivate an ansatz for the PsP and then adjust the free parameters in the ansatz to desired observables as, for example, atomic spectra or asymptotic profiles of electronic wave functions.

3.3.1.2 Practical Pseudopotentials

It is a long way from these still exact grounds to a practical, efficient, and reliable PsP. There is also a broad range of choices derived from more or less strict paths. Formally, the simplest choice are the local PsPs [86] and soft shapes are preferable for numerical handling. In this case, the PsP is a "simple" function of the position, and the same for all electrons, for a given ion. Such local PsPs work in simple systems as typically alkaline metals in which the valence electron is unique and well separated from core states. But in many other cases such as organic atoms, local PsPs have to account for more valence electrons (e.g., four in a carbon atom), and one constructs "nonlocal" PsP which also depend on the electronic angular momentum L with respect to the ion. This type of PsP is much more involved but also more flexible.

Figure 3.5 demonstrates the effect of PsPs on the description of the wave functions. It compares the exact wave function for the 3s state ($L = 0$) in the Ar atom (obtained from an all-electron calculation) with the corresponding wave function as obtained from a nonlocal PsP. The nodal structures in the interior of the atom are much different. But this is acceptable because it is unimportant for the description of long-range features and in particular bonding (see Section 1.1.3.2). What counts is that the two wave functions nicely agree outside the region of the core states, above 1 a_0. This is visibly well fulfilled here, and this should be the case for any

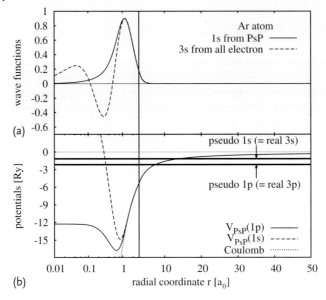

Figure 3.5 Illustration of pseudopotentials (PsP) for an Ar atom. An horizontal logarithmic scale is used below 3 a_0, and a linear one above. Illustrated is the effect of the PSP on the wave function (a), with the exact wave function of the 3s state from an all-electron calculation with a self-interaction corrected LDA (dashed line); and ground state in a non-local PsP (solid line). Also shown are the actual PsPs (b) for the "pseudo" 1s state (dashed curve) and for the 1p state (solid curve), compared with the mere Coulomb potential of the Ar^{8+} ion (faint dotted line). The two horizontal lines indicate the single-particle energies for the "pseudo" 1s and 1p states, and coincide with those of the "real" 3s and 3p states from an all-electron calculation.

well-fitted PsP. For completeness, Figure 3.5b displays the PsPs for the "pseudo" 1s and 1p states. The 1s state has $L = 0$ and is equivalent to the true 3s in the all-electron atom, while 1p has $L = 1$ and is equivalent to the true 3p. Both potentials asymptotically match the ionic Coulomb potential, but strongly differ in the interior. In particular, the Coulomb singularity at ionic site is removed by the smooth PsP. Note that the $L = 0$ potential (see dashes) has a larger repulsive core than that for the $L = 1$ case (see solid line). It simulates the strong Pauli repulsion from the $L = 0$ core states.

3.3.1.3 Transferability of the Pseudopotentials

The correct asymptotics of the pseudo wave functions is one crucial condition which is closely related to the another crucial demand of transferability, namely the fact that one and the same PsP is appropriate to describe molecular binding up to the bulk. This requires that an atomic calculation with the PsP reproduces as well as possible the energy levels of the states near the Fermi surface, including the first excited states. Most widely used are fitting procedures which rely on atomic calculations. Starting from reliable all-electron calculations for the atom, including several excited states, one selects the relevant atomic states (valence and a few excited states) to be described by the PsP; one chooses an appropriate form for the PsP, and then one adjusts its free parameters to optimally accommodate the chosen states. A careful fit to sufficiently many atomic states automatically arranges the polarizability, thus yielding correct binding properties from molecule to bulk. Beyond reproducing electronic single-particle energies, key constraints in the fit can also be an exact matching of the (possibly nodeless) pseudo wave functions with the atomic wave functions outside a given core region (see Figure 3.5a), which ensures proper screening behavior.

3.3.1.4 Core Polarization Effects

Thus far, the motivation for a PsP employed the working assumption that the eliminated core states are perfectly inert. At second glance, one realizes that an external charge applies a Coulomb force on the core which, in turn, reacts with a small polarization. The polarization effect lowers the energy of the combined system and this contributes to the effective potential. Those polarization potentials are a leading ingredient in effective rare gas atom–atom potentials, for example, for Ar–Ar [87]. They also add to the electron–ion PsP. In most cases, they are implicitly contained in the adjustment of the effective PsP. One has to keep in mind that such an adjustment then fixes the static polarization properties. An explicit handling of dynamical polarization is advisable if the frequency of core vibrations is not sufficiently well separated from the dynamical processes under consideration. An example is the description of noble metals with only one valence electron per atom, for instance in Cu, where the d shell is not so far away from the 4s shell (see Figure 1.4). The same holds in Ag and Au, commonly used materials in cluster physics. By far, most PsP do not count d electrons as core electrons in such systems, but require an explicit treatment. If one deliberately wants to eliminate them, one should thus treat their dynamical polarization explicitly. Only this guarantees an appropriate description of the optical response of noble metal clusters.

3.3.2
Jellium Approach to the Ionic Background

3.3.2.1 The Original Jellium

Simple metals have valence electrons with long mean free path throughout. The relevant electrons move with momenta near Fermi momentum, which corresponds to a spatial resolution of the order of the Wigner–Seitz radius r_s. The fine details of the ionic background are thus seen by the electrons only in an average manner. This motivates the jellium approximation in which the ionic background is smeared out to a constant positive background charge. It is a standard approach in bulk metals [10]. The generalization to finite clusters is straightforward. In its simplest form, one carves from the bulk a finite, homogeneously and positively charged sphere of radius $R = r_s N_{\text{ion}}^{1/3}$, whose total ion charge reproduces the given ionic charge $e N_{\text{ion}}$. This steep spherical jellium model serves very well to outline the basic properties of metal clusters, in particular concerning gross shell structure (see Section 4.1.1.1).

3.3.2.2 The More Flexible Soft Jellium

A more flexible approach is achieved when allowing for a finite surface width and for some deformation, yielding the soft (deformed) jellium model:

$$\rho_{\text{jel}}(r) = \frac{3}{4\pi r_s^3} \left[1 + \exp\left(\frac{|r| - R(\vartheta, \phi)}{\sigma_{\text{jel}}} \right) \right]^{-1}, \quad (3.13\text{a})$$

$$R(\vartheta, \phi) = R_{\text{jel}} \left(1 + \sum_{lm} \alpha_{lm} Y_{lm}(\vartheta, \phi) \right), \quad (3.13\text{b})$$

with R_{jel} defined by normalization to total particle number: $\int d^3 r \rho_{\text{jel}} = N_{\text{ion}}$. The central density is determined by the bulk density $\rho_0 = 3/(4\pi r_s^3)$. The parameter σ_{jel} accounts for a smooth surface transition from ρ_0 to zero. To give an impression about the surface width, the transition from 90 to 10% bulk density is achieved within $4\sigma_{\text{jel}}$. Deformations are parametrized in $R(\vartheta, \phi)$ through the coefficients in front of the spherical harmonics Y_{lm} with the parameters α_{lm} determining the amount of deformation. For example, axially symmetric ellipsoids are tuned by α_{20}, positive values producing prolate (cigar-like) and negative ones oblate (pancake-like) shapes. Octupole (pear-like) shapes are generated by α_{30} and can have a considerable influence in metal cluster spectroscopy. Next come hexadecapoles α_{40}, which play a role for fine-tuning. Higher moments are rarely considered. Triaxial shapes are produced by moments with $m \neq 0$. This rich palette of shapes suffices for most applications. Some jellium parameters are universal for a given material. For example, typical values are $r_s \sim 4a_0$ and $\sigma_{\text{jel}} \sim 0.9a_0$ for Na clusters, $r_s \sim 2.66a_0$ and $\sigma_{\text{jel}} \sim 0.76a_0$ for Mg clusters, or $r_s \sim 3a_0$ and $\sigma_{\text{jel}} \sim 0.78a_0$ for Ag clusters. The deformation α_{lm} is determined by electronic shell effects and sensitively depends on the actual electron number N_{el} (see Section 4.1.2).

Figure 3.6 compares the soft jellium model to an explicit ionic structure in terms of ionic background density, electron density, and Kohn–Sham potential (see Section 3.2.4) for the (nearly) spherical cluster Na_{41}^+. Ionic densities are not point densities but smooth densities such that they produce the ionic pseudopotentials by solving the ionic Coulomb problem. The background densities look at first glance much different. Note, however, that the plot shows a cut along the z axis in Figure 3.6b. Top and bottom ions lie exactly on this axis and these two ions show here a dramatic peak each. The smaller peaks next to the large ones are traces of ions which are slightly off, but still close to the axis. The jellium density looks, of course, much different, although it is crucial that it covers about the same range. The electronic densities are comparatively similar in spite of the dramatically different ionic densities. This shows that the electron cloud of a metal cluster is not so sensitive to the details of the ionic background, so that the jellium model is appropriate for describing several electronic properties (as shell effects, optical response). This conclusion is corroborated by the Kohn–Sham potentials shown in

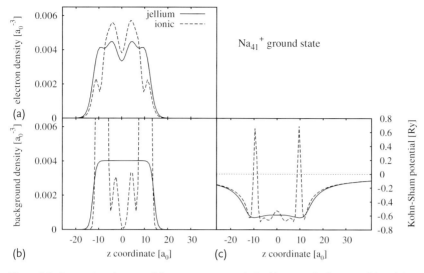

Figure 3.6 Comparison between fully ionic structure (dashed lines) and jellium model (solid lines) for ground-state densities and potentials of Na_{41}^+, showing electron densities (a), ionic background densities (b) and total Kohn–Sham potentials (c) (see Section 3.2.4).

Figure 3.6c. Besides the two narrow ionic cusps, both potentials are globally similar and thus produce similar electronic solutions.

3.3.2.3 From Pseudopotentials to Jellium

The soft jellium approach can be deduced from the concept of PsP. Let us identify the steep jellium as resulting from smearing a collection of point charges to a smooth background. We now associate a local PsP to each point charge. A soft jellium then emerges as a folding of the steep jellium with the given PsP. This definition of smoothed ionic background yields a perfect description of magic shell closures and optical response in metal clusters [88], and that without any free parameter (r_s is taken from bulk and the surface profile comes from the PsP). The same strategy of folding steep jellium with a PsP can be pursued for a nonlocal PsP. This yields effective Hamiltonians with nonlocal contributions, such as coordinate-dependent effective mass or angular momentum-dependent potentials [89]. The thus generalized models are also often called pseudojellium, or pseudo-Hamiltonians. These nonlocal models extend the range of applicability of the simple jellium to a broader range of metals which require nonlocal PsP, as for example Li.

3.4
Observables

3.4.1
Energies

The total binding energy is naturally the most prominent observable in theoretical calculations. However, it is not easily accessible in experiments. Measurements are more likely to provide differences of energies. One is the monomer separation energy as the adiabatic energy difference $E_{\text{mon}} = E(N_{\text{el}}, \mathbf{R}^{(N_{\text{ion}})}) - E(N_{\text{el}}-1, \mathbf{R}^{(N_{\text{ion}}-1)})$, where the reference energies are to be taken from fully relaxed ionic configurations. Another energy difference is the vertical ionization potential (IP) $E_{\text{IP}} = E(N_{\text{el}}, \mathbf{R}^{(N_{\text{ion}})}) - E(N_{\text{el}}-1, \mathbf{R}^{(N_{\text{ion}})})$, where the electronic state in the $N_{\text{el}} - 1$ system has relaxed but the ions are kept in their original configuration.

As a byproduct of mean-field calculations, one also obtains the series of single-electron energies ε_i for the occupied states. DFT does not guarantee that these are correctly described. And indeed, the mere LDA underestimates the IP when identified with the energy of the HOMO level. The self-interaction correction (see end of Section 3.2.4) makes single-electron energies significantly more reliable. The latter energies can be observed experimentally by photoelectron spectroscopy, see Sections 3.4.5 and 2.3.4.

3.4.2
Shapes

Clusters cover two subsystems, ions and electrons, each one carrying its own shape. At a detailed level, the ionic shape is given as a point distribution by a set of coordinates $\{\mathbf{R}_I\}$ while the electron cloud is associated with a smooth density distribution $\rho(\mathbf{r})$. Ionic structures are often characterized in terms of point symmetries, that is molecular ones for small and crystal ones for large clusters. Comparison of different systems or establishing trends requires to condense the rich information in $\{\mathbf{R}_I\}$ and $\rho(\mathbf{r})$ to a few key numbers. A useful measure is provided here by global shape parameters in terms of multipole moments. These are generally expectation values of certain local functions $f(\mathbf{r})$ where $\langle f \rangle = \int d^3 r \rho(\mathbf{r}) f(\mathbf{r})$ for electrons and $\langle f \rangle = \sum_I f(\mathbf{R}_I)$ for ions. The leading quantity is the monopole moment, that is the root-mean-square radius $r_{\text{rms}} = \sqrt{\langle r^2 \rangle / N}$ where N is the number of electrons for electronic moments and the number of ions for ionic moments. Next important are the electronic dipole, $\mathbf{D} = \langle \mathbf{r} \rangle$, and quadrupole momenta, $Q_{ij} = \langle r_i r_j - \delta_{ij} \frac{1}{3} r_{\text{rms}}^2 \rangle$, $(i, j = x, y, z)$. The electronic dipole is the key observable in electronic dynamical response (see Section 3.4.4), while the ionic dipole is practically the center-of-mass of the cluster.

The shape in the sense of deformation is quantified by the quadrupole momenta. There is the Cartesian quadrupole Q_{ij}, or equivalently the spherical quadrupole moment $Q_{lm} = \langle r^2 Y_{2m} \rangle$ ($m = 0, \pm 1, \pm 2$). More relevant are the dimensionless quadrupole momenta

$$a_{2m} = \frac{4\pi}{5} \frac{Q_{lm}}{N r_{\text{rms}}^2}, \quad Q_{lm} = \langle r^l Y_{lm} \rangle. \tag{3.14}$$

The total quadrupole deformation is given by $\beta_2 = \sqrt{\sum_m a_{2m}^2}$. For a geometrical interpretation, one should evaluate a_{2m} in the system's principal axes in which there remain only two relevant deformations, namely a_{20} and $a_{22} = a_{2,-2}$. Values $a_{22} \neq 0$ stand for triaxial systems, while $a_{22} = 0$ signals axial symmetry. The dimensionless momenta are free from the scale of the overall radius and therefore are immediately related to the geometry of a cluster.

The quadrupole deformation is often alternatively characterized by the moments of inertia $I_{ii} = \langle r^2 - r_i^2 \rangle$, $i = x, y, z$, to be evaluated in the frame of principal axes of the cluster (this is the geometrical definition; true physical inertia are obtained by multiplying that with the particle mass). These moments of inertia are directly related to the above moments. Quadrupole moments and, equivalently, moments of inertia are in principle different for ions and electrons. But the Coulomb force tries to keep ionic and electronic moments close to each other (at least, for the low l moments discussed here). There is thus a close relation between moments characterizing cluster shape and the Mie plasmon response in metal clusters, measured by optical response, see Figure 2.7 and also Section 4.2.

3.4.3
Stationary Response: Polarizability and Conductivity

The static dipole polarizability α_D is a key observable in atomic and molecular physics, thus for clusters as well. The computation is straightforward. One applies a static external dipole field $U_{\text{ext}}(r) = e E_0 \cdot r$ and performs static calculations for various values of $E_0 = |E_0|$. One then obtains a dipole moment $D = e \langle r \rangle = D(E_0)$ and from that, the dipole polarizability [27]

$$(\alpha_D)_{ij} = \frac{\partial D_i}{\partial E_{0,j}}\bigg|_{E_0=0}$$
$$= 2e^2 \sum_n \frac{\langle \Psi_0 | r_i | \Psi_n \rangle \langle \Psi_n | r_j | \Psi_0 \rangle}{E_n - E_0}, \quad \{i, j\} \in \{x, y, z\}. \tag{3.15}$$

The first expression requires only static calculations. The second (equivalent) expression computes dipole response via the excitation spectrum where $n = 0$ is the ground state and $n > 0$ labels the excited states with associated energies E_n.

A microscopic evaluation of electrical conductivity is, on the contrary, cumbersome. There exist several such attempts for clusters, see [90] for a review. The basic step is the same as in macroscopic physics, namely to consider the response current to an applied voltage in the linear regime. This amounts to computing the electron transmission function from one contact to another, which can be done by standard many-body techniques. The particular problem with computing conductivity at the nanoscale is the contact which is an integral part of the setup, and therefore sensitively influences the resulting transport properties. Careful modeling of the contacts is thus required and represents a demanding task, similar to the problem of clusters in contact with a substrate (see Section 5.2).

3.4.4
Linear Response: Optical Absorption Spectra

Optical response is a key observable in cluster physics, as already discussed in Section 2.3.3. We now want to briefly discuss how to compute such a quantity. Since clusters are small as compared to the wavelength of visible light, they couple to photons only by the dipole operator. Optical absorption strength is thus closely related to the distribution of dipole strength. All methods to describe electrons can be used to compute the spectrum of dipole excitations. Conceptually most straightforward (but technically most involved) is CI which computes the excitation spectrum on the same footing as the ground state. This is however limited to small clusters because in practice, a reliable computation of excitations requires an even larger basis than what would be sufficient for mere ground-state calculations (see Figure 3.3).

The next microscopic approach to optical response relies on TDLDA (Section 3.2.4). This can be done in two ways. The first one is to linearize the TDLDA equations, which delivers a secular equation for a coupled-oscillator problem. Linearized TDLDA is one of the most basic schemes in many-body physics, known under different names, such as linear response theory or random-phase approximation (RPA) [28]. RPA is advantageous in cases with high symmetries. It provides discrete excitation energies E_n and wave functions Ψ_n. The strength distribution of the dipole in the x, y, or z direction is then obtained from the corresponding transition strengths, augmented with some Lorentzian smoothing

$$S_{D_i}(E) = \sum_n |\langle \Psi_n | \hat{D}_i | \Psi_0 \rangle|^2 \frac{\Gamma/\pi}{(E - E_n)^2 + \Gamma^2} , \qquad (3.16)$$

where $i \in \{x, y, z\}$. The photoabsorption strength $\sigma_{\text{abs}}(E)$ is computed the same way, but with extra energy weights E_n. The width Γ simulates various broadening mechanisms beyond RPA accounting for electron escape, electron–electron collisions and thermal shape fluctuations. RPA can become extremely cumbersome for systems where all symmetries are broken.

The second computation of dipole spectra uses full TDLDA and explores the response of the system to laser pulses. The response spectrum in the linear regime can be computed by using an extremely short pulse probing all frequencies at once. The ideally short pulse is modeled by an instantaneous (small) boost applied to each Kohn–Sham ground-state orbital as $\varphi_{\text{gs},i} \to \varphi_i(\mathbf{r}, t = 0) = \exp(i\mathbf{p}_{\text{boost}} \cdot \mathbf{r})\varphi_{\text{gs},i}$, where $\mathbf{p}_{\text{boost}}$ has the dimension of a momentum and tunes the amplitude of excitation. To evaluate the response, one propagates the system by TDLDA propagation. This delivers the density $\rho(\mathbf{r}, t)$ from which one samples a protocol of the dipole moment as $\mathbf{D}(t) = \int d^3 r \mathbf{r} \rho(\mathbf{r}, t)$. After a sufficient simulation time T_{max}, one Fourier transforms the dipole signal into the frequency domain and finally obtains the

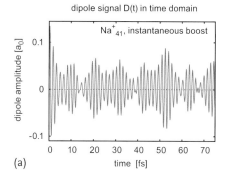 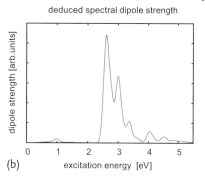

Figure 3.7 Example of the computation of spectral distributions in the test case Na_{41}^+. (a) Time evolution of the dipole momentum $D_z(t)$ in the z direction after an instantaneous boost. (b) Dipole strength distribution obtained by Fourier transforming $D_z(t)$.

dipole strength as

$$S_{D_i}(E) = \Im\{\widetilde{D}_i(\omega)\}, \quad \widetilde{D}_i(\omega) = \int dt e^{i\omega t} D_i(t), \quad \omega = \frac{E}{\hbar}, \quad (3.17)$$

with $i \in \{x, y, z\}$. The spectral resolution is given by $\delta\omega = 2\pi/T_{\max}$. The response of the system then contains a relevant picture of the dipole strength distribution. In practice, however, one should perform spectral filtering to avoid artifacts due to the fact that the signal $D_i(t)$ has usually not relaxed to zero at the end of the analyzing time [91]. For more details on spectral analysis, see [46] and references therein. We illustrate the scheme in Figure 3.7 for Na_{41}^+. Figure 3.7a shows a part of the time evolution of the dipole signal in the z direction, and Figure 3.7b the corresponding dipole strength (obtained with Gaussian filtering of width 15 fs to smooth the pattern). The dipole signal $D_z(t)$ shows regular oscillations with the dominant frequency of about 2.8 eV with considerable beating. The dominant frequency correlates to one large peak in the spectral strength, and the beating to the broadening of the peak by spectral fragmentation.

The same spectral analysis can be performed with other observables as, for example, higher multipoles or spin modes. It can also be applied to the computation of ionic vibration spectra, however, at much longer time scales. Furthermore, full TDLDA allows to go beyond the linear regime (by a large boost or a strong laser excitation). The more appropriate observable in such cases is the power spectrum $\mathcal{P}_{D_i}(\omega) = |\widetilde{D}_i(\omega)|^2$. Spectral analysis via full TDLDA furthermore allows to incorporate the effect of ionic motion on the electronic spectra by simply using the coupled ionic and electronic dynamics of TDLDA-MD (see Section 3.1.3.1). Mie plasmon oscillations are much faster than ionic motion which allows to consider instantaneous snapshots for a fixed ionic configuration (and this is what is usually done). There is, however, an important effect from ionic motion leading to line broadening (see Section 4.2.3).

3.4.5
Electron Emission

Electron emission delivers several observables. At a global level, one can consider the total number of emitted electrons (average ionization) and the ionization probabilities for each final charge state separately. More detailed information is contained in the angular distribution of the emitted electrons and in their kinetic energy spectrum (see Section 2.3.4). At the production side, there exist two different mechanisms for electron emission: thermal evaporation and direct emission.

3.4.5.1 A Quick Glance at Thermal Emission

The thermal emission mechanism is conceptually simple. Excitation is first converted into thermal energy of the electron cloud. It then happens once in a while that one electron gathers sufficient kinetic energy to surmount the emission threshold. For most purposes the simple estimate suffices with the Weisskopf formula [92]:

$$W_{\text{evap}} = \frac{2\sigma_{\text{capt}}}{\hbar^3 \pi^2} m T^2 \exp\left(-\frac{E_{\text{IP}}}{T}\right), \tag{3.18}$$

where T is the actual temperature, E_{IP} the energy of the ionization potential, and σ_{capt} the cross-section for electron capture. The latter one may be approximated by the geometric cross-section ($\sigma_{\text{capt}} \sim \pi R^2$, R being the system's radius). The resulting relaxation times W_{evap}^{-1} are very long at room temperature (from ns up to μs), but they can shrink to the ps, even fs, range for high excitations with temperatures of 1000 K or more (see Figure 1.3). Thermal emission yields isotropic photoelectron angular distributions (PAD) and smooth exponentially decreasing photoelectron spectra (PES) whose slope provides an estimate of the temperature T.

3.4.5.2 Direct Electron Emission

Direct emission occurs if an electron immediately follows the force induced by an external electrical field (laser or bypassing ion), possibly amplified by response fields of the cluster. Strong fields lower the barrier so that the electron can directly overcome it. Less strong fields can still trigger direct emission by tunneling (see Figure 4.15). Direct emission exhibits a strong anisotropy of the PAD (see Figure 4.13) and very structured PES (see Section 4.3.2). It thus delivers rich information on cluster structure and dynamics. But it competes with the less instructive thermal emission. It prevails in the first time steps, whereas thermal emission takes over on longer time scales. Short laser pulses are thus favorable to preferably induce direct emission.

Calculations of direct emission require a time-dependent method to track the dynamical process in detail, and a numerical representation which properly deals with continuum states. An efficient approach is once again TDLDA for the time evolution, for example using a coordinate-space grid representation combined with absorbing boundary conditions [46]. This allows for true electron emission since it formally means that each single-electron wave function on the grid gradually loses norm related to the outflow of electrons from the corresponding state. Moreover, efficient absorbing bounds will suppress electron reflection from the boundaries of the numerical box which means, in turn, that there is practically no inflow of electrons near the absorbing zone. These features are used to compute various emission observables, from the net ionization to detailed PES and PAD, the latter ones being quite involved to compute. Practical examples of PES and PAD will be given in Section 4.3.

4
Some Properties of Free Clusters

A typical experiment consists of three steps: preparation, excitation, and observation. The theoretical modeling can be more or less divided into the same steps. We have explained in Chapter 3 how to prepare the ground state of a cluster by stationary solution of the electronic problem and optimization of ionic structure. We have also outlined all the tools to describe the dynamic evolution. But before we can extract observables from the dynamical simulations, we need to specify the modeling of excitation mechanisms and the handling of observables. That will be done in the first two sections of this chapter. The subsequent sections then present a brief tour through the basic properties of clusters concerning structure and optical response.

4.1
Ionic and Electronic Structure

Cluster structure predominantly manifests itself in the binding energy. Thus the binding energies, trends thereof, and energy differences provide basic information on the underlying ground-state structure. This will be exemplified in Section 4.1.1 for the case of shell closures and associated magic numbers. Shell structure also has a strong impact on the global deformation. This aspect will be discussed in Section 4.1.2. Experimental information on deformation is gained for metal clusters by optical response, as will be discussed in Section 4.2.2. A direct measurement of the spatial cluster structure by scattering experiments as done, for example, in crystallography is very cumbersome and rarely done. We will not address this aspect here in depth.

4.1.1
Magic Numbers

4.1.1.1 Electronic Shells

Figure 4.1 shows trends of energies for neutral Na clusters, for abundances and for IP. The abundances show peaks and the IP steps at mass numbers $N = 8, 20, 40$, and 58. It can be associated to electronic shell closures for spherical shapes, here in the Kohn–Sham potential of Na clusters. For an illustration, see Figure 3.4 which

An Introduction to Cluster Science, First Edition. Phuong Mai Dinh, Paul-Gerhard Reinhard, and Eric Suraud.
© 2014 WILEY-VCH Verlag GmbH & Co. KGaA. Published 2014 by WILEY-VCH Verlag GmbH & Co. KGaA.

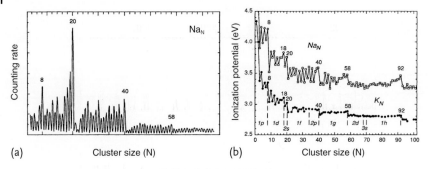

Figure 4.1 Signatures of electronic shells in Na clusters for abundances of Na$_N$ in an evaporative ensemble (a) and ionization potentials of Na and K clusters (b). Composed from [17].

depicts the Kohn–Sham potential and the level sequence for the Na$_{58}$ cluster. One recognizes the shell closures at $N = 8, 20, 40, 58$ and 92. The effect is the same as for rare gases amongst the atoms. The binding is strongest if the number of electrons matches a filled electronic shell. For clusters, the strong binding makes the systems with magic electron number particularly stable in an evaporative ensemble which, in turn, yields the peaks in the abundances. Corroborating information on shell closures is provided by the systematics of IP, see Figure 4.1b. Electrons are particularly well bound at shell closure and only loosely just above. This leads to a large step in the IP which thus signals an electronic shell closure. The results from the IP fully agree with the shells seen in the abundances.

A similar reasoning applies for the dissociation energy $D(N) = E(N) - E(N-1)$, that is the energy required to separate one monomer from a cluster of N atoms. In fact, this is close to the quantity which determines the relative abundance in a thermal ensemble, namely the difference of free energies $F(N) - F(N-1)$ (where $F = E - TS$) which directly enters the Boltzmann factor in the equilibrium distribution. A better visualization is given by the difference of dissociation energies, that is by the second difference of free energies $\Delta_2 F = F(N+1) - 2F(N) + F(N-1)$. Large values of $\Delta_2 F$ signal large abundances. Theoretical results for $\Delta_2 F$ show peaks in nice agreement with experimental data [30].

4.1.1.2 Shells and Supershells

The pioneering measurements of [5], as reproduced in Figure 4.1a, discovered the magic shells up to $N = 92$. This has triggered a long search for ever larger clusters and their magic shells in a combined effort. A summary is shown in Figure 4.2. Note that abundances drop over several orders of magnitude when going up so far in particle number. A scaling factor has thus been applied to counterweight the average trend and to allow one to draw all results on the same scale. The shell closures are distinguished here by deep negative spikes. The associated magic electron numbers are indicated. A particularly interesting feature is the trend for the amplitude of the shell effects. It fades away for increasing cluster size up to a minimum (for around $N \sim 600$–800) and returns again when stepping further up ($N \gtrsim 1000$). This beating of the shell gap over the sequence of magic numbers

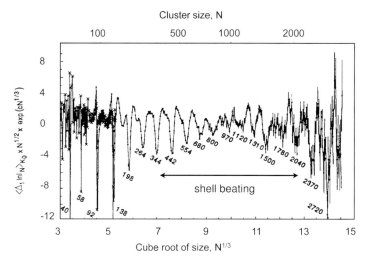

Figure 4.2 First difference of logarithmic experimental yields I_N, augmented by a scale factor, $\sqrt{N}\exp(cN^{1/3})$, to equilibrate the average trends, as a function of cluster size N. Magic electron numbers (shell closures) are associated with each negative spike as indicated. The amplitude of the shell effects shows a minimum around $N \approx 1000$ associated with a beating between two different sequences. Adapted from [30].

is called supershells. The effect was predicted for cavity modes long ago [93] and metal clusters have provided the first quantum system to experimentally display this effect [69]. The mechanism for producing supershells is understood in a semiclassical picture of shell quantization (periodic orbit theory). They emerge from a competition of the two dominant periodic orbits, those forming a triangle versus those forming rectangles. For a detailed discussion, see [94].

4.1.1.3 Ionic Shells

It is also plausible that the geometric configuration of atoms plays a role in cluster binding and one expects atomic shell closures for particularly favorable geometries (minimal missing links). The case of atomic shells has been extensively discussed in the review [95]. Figure 4.3 shows an example for large Ca clusters with icosahedral symmetry. Differences of abundance are plotted to work out the effect more clearly. Clear peaks are indeed seen, corresponding to atomic shell closures due to completion of a full icosahedral shape. Each (although small) oscillation corresponds to the completion of subfaces. Note that atomic shells sensitively depend on the material, more than electronic shells do. There are at least as many shapes and shell sequences as there are crystal symmetries.

One has thus to be aware of two different sorts of shell closures: atomic and electronic ones. Electronic shells become apparent in metal clusters at large temperatures (around the melting point), while atomic shells are best seen at very low temperatures where thermal shape fluctuations are small. Due to their different nature, atomic and electronic shells scale very differently, as we can see from comparing Figures 4.2 and 4.3. The magic electron numbers come along in a much

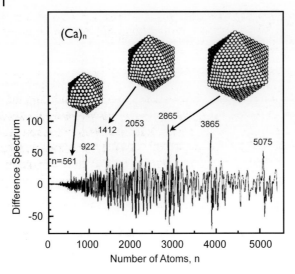

Figure 4.3 Difference of mass abundances for Ca clusters. Some shapes associated with pronounced atomic shell closures are shown and the link is indicated by an arrow. After [95].

denser sequence than the atomic ones because it is much more demanding to arrange a regular piece of a crystal with clean flat surfaces all around.

4.1.2
Shell Structure and Deformation

Electronic shell effects play a crucial role for the global shape of metal clusters. They constitute a typical realization of the Jahn–Teller effect [96]. Indeed the electronic ground state does not like to be degenerated. Molecules (and clusters) thus accommodate their shape to reach an unambiguous electronic ground state. This is demonstrated in Figure 4.4 for small metal clusters using a deformed harmonic oscillator shell model, the Clemenger–Nilsson model [17]. Single-electron levels are drawn as functions of the (axially symmetric) quadrupole deformation α_{20}, see Eq. (3.14). Each level can carry two electrons (spin up and spin down). The configuration for a given electron number is found by filling the levels from below according to the Pauli principle. The labels in the plot indicate the electron number reached at a given energy and deformation. At a spherical shape ($\alpha_{20} = 0$), we recognize again the magic electron numbers 2, 8, and 20 (see Figure 4.1). It is obviously impossible to obtain other electron numbers at $\alpha_{20} = 0$ without running into degeneracies. Different closures and gaps open up at different deformations and the cluster moves to those deformations where it finds a nice gap at given N_{el}. We can read off from the figure that $N_{el} = 4$ and 10 are prolate (negative α_{20}), while $N_{el} = 6$, 14 and 18 are oblate (positive α_{20}). One misses the cases $N_{el} = 12$ and 16. These numbers do not fit into the present scheme (the point marked 16' turns out to be only an isomeric state) based on axial symmetry. One needs to deal

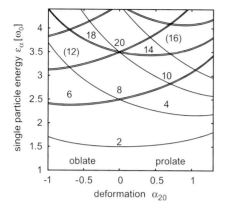

Figure 4.4 Electronic shell sequence for the axially symmetric, deformed harmonic oscillator [17, 28] drawn versus the deformation parameter α_{20} as defined in Eq. (3.14). Numbers in the shell gaps indicate the numbers of electrons reached when filling up to the last level below that number. The cases $N = 16$ and $N = 12$ are given in brackets because both these systems actually prefer a triaxial deformation.

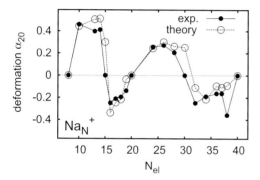

Figure 4.5 Predictions of the ground-state deformation parameter α_{20} as a function of the number of electrons N_{el} in cationic Na_N clusters ($N_{el} = N - 1$), as deduced from the Clemenger–Nilsson model, and compared with experimental data from [34].

with fully triaxial shapes to create space for a gap at $N_{el} = 12$ and 16. However, this symmetry breaking is not as efficient as the first step from spherical to axial symmetry. As a consequence, the HOMO–LUMO gap for triaxial shapes is small and these clusters are not so well bound.

The shape of small metal clusters is experimentally assessed through the splitting of the Mie plasmon resonance as will be explained in Section 4.2.2. Figure 4.5 compares the measured sequence of cluster deformations with those deduced from the Clemenger–Nilsson model. The agreement is impressive in view of the simplicity of the model. One sees the spherical shape for the magic electron numbers (2, 8, 20, 40). Substantial deformation develops mid-shell between the magic numbers. Prolate shapes prevail just above magic numbers and oblate ones just below. The transition from the prolate to the oblate regime should be accompanied by triaxial shapes. However, their occurrence is not so clearly discriminated by the experimental data. It is actually extremely hard to disentangle the splitting into three different plasmon peaks in the data. There also exist full Kohn–Sham calculations (see Section 3.2.4) including detailed ionic structure for this sequence. They corroborate the above findings. The electronic shells in metal clusters is indeed the leading agent in determining ionic structure.

4.2
Optical Response

Optical absorption strength is a key tool to investigate structure and dynamics of quantum systems. The detection of discrete absorption lines in atomic hydrogen was one of the cornerstones of quantum mechanics. The pattern of optical absorption provides crucial clues on the structure of molecules; corresponding selection rules often allow one to assess the underlying point symmetry. Optical absorption also plays a key role in cluster physics. We will discuss in this section a few typical examples. There is an emphasis on the case of metal clusters because these follow very general patterns due to the dominance of the Mie plasmon resonance (see Section 2.2.3). Nonmetallic materials have more specific spectra which can only be discussed case by case.

4.2.1
Basic Features

Figure 4.6 shows optical absorption spectra for clusters of two different materials, one with covalent binding (Figure 4.6a) and one with metal binding (Figure 4.6b). The spectrum in Figure 4.6a is distributed over several comparable peaks and extends over a broad range of energies. This is a typical situation in many clusters with covalent, ionic or van der Waals binding. Little can be said in general because the actual spectrum very much depends on the details of ionic and electronic structure. This is different for metals as can be seen in Figure 4.6b. The spectrum is dominated by one well-concentrated peak, the Mie plasmon resonance. The picture is similar for all (simple) metal clusters. The frequency resides for all cluster sizes in the same range as given by the Mie frequency (2.6), which follows the trend $\omega_{\mathrm{Mie}} \propto r_s^{-3/2}$ (see Section 2.2.3). There are small, but informative, variations in the shape and width of the resonance width. All features can be explained and predicted on very general grounds as will be discussed in the following.

The Mie plasmon frequency is obtained by a simple collective model, see Section 2.2.3. A cluster is actually a many-electron system. One wonders how a collective mode could emerge from a puzzling manifold of single-electron excitations. We try to explain this in the example of the cluster $\mathrm{Na}_{93}{}^+$ with Figure 4.7. The simplest concept of excitation is that one particle leaves an occupied state h, thus cutting a hole therein, and fills a formerly unoccupied state p. This is called a one-particle-one-hole (1ph) state whose excitation energy is simply taken as the differ-

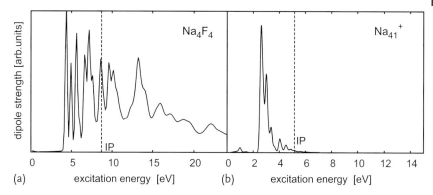

Figure 4.6 Photoabsorption spectra as calculated from TDLDA (see Section 3.2.4) for a cluster with ionic binding (a) and a metal cluster (b). The IP are indicated by vertical dashed lines in both cases.

ence of single-particle energies, $\varepsilon_p - \varepsilon_h$. Figure 4.7d shows the sequence of the 1ph transition energies. Each vertical line stands for one excitation. The spectral density is large and one could expect a fuzzy spectrum. Figure 4.7c shows the dipole strength from such pure 1ph excitations, that is the 1ph spectrum weighted by the squared dipole matrix element. Only a few lowest-energy transitions around 1 eV carry significant dipole strength. The huge variety of 1ph states with small dipole strength at higher energies is not visible in the dipole strength. The full TDLDA spectrum is shown in Figure 4.7b. The now accounted for residual interaction produces a significant up-shift of dipole strength, basically to the Mie plasmon frequency as indicated by a vertical dashed arrow. The key ingredient for the up-shift is the strongly repulsive Coulomb interaction between the 1ph states. The essence of the shift is already correctly described in the Mie model where the spring constant for the oscillating electron cloud is determined by the Coulomb force. However, the collective resonance peak resides in a region of high density of 1ph states (Figure 4.7d). Thus the collective dipole strength is redistributed over all the energetically close 1ph states. The result is a somewhat fragmented resonance peak. This is called Landau fragmentation in analogy to Landau damping in the case of a continuous electron gas [8].

The surface plasmon resonance in large Na clusters thus already carries some width due to Landau fragmentation. This is obviously not enough to deliver the smooth experimental pattern shown in Figure 4.7a (dashed line). The point is that this measurement was performed at a rather high temperature of 400 K. This causes fluctuations of the cluster shape at the pace of ionic motion (since 400 K are almost 0 for electrons), thus on a time scale much longer than the typical period of electron oscillations. The resonance changes with shape and, because of the very different time scales involved, the various spectra incoherently add up. This brings another piece to the width, see Section 4.2.3. Here we simulate this thermal broadening by folding the TDLDA result of Figure 4.7b with a Gaussian of full width at half maximum of 0.14 eV [98]. The result, shown in Figure 4.7a, agrees very nicely with the experimental spectrum. Figure 4.7 thus demonstrates the vari-

4 Some Properties of Free Clusters

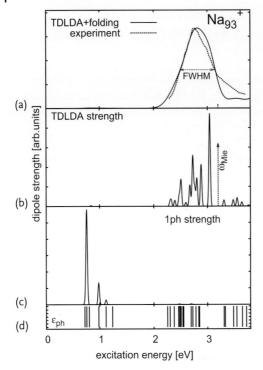

Figure 4.7 Demonstration of the coherent collective shift in the dipole excitation spectra for the cluster Na_{93}^+, computed with DFT (see Section 3.2.4) and the spherical jellium model (3.13a)–(3.13b) for the ionic background charge ($r_s = 3.9\, a_0$, $\sigma = 0.9\, a_0$). The first spectrum shows a TDLDA spectrum (a) folded with a Gaussian of width 0.14 eV to simulate thermal averaging (solid line) compared with the experimental result from [97] (dashed). The second shows the dipole strength from full TDLDA, where the surface Mie plasmon from Eq. (2.6) is indicated by the arrow (b). Next shows the dipole transition strength (optical absorption) in a pure 1ph picture (c). Finally shown are the pure one-particle-one-hole (1ph) energies $\varepsilon_{ph} = \varepsilon_p - \varepsilon_h$ (d).

ous contributions from which the optical absorption strength is composed. TDLDA embraces the dominant (collective) effect of the Coulomb repulsion and all details of the 1ph structure. Thermal effects have to be added in terms of an incoherent superposition of TDLDA results at various shapes.

4.2.2
Impact of Deformation

The example Na_{93}^+ dealt with an approximately spherical clusters whose spectrum does not change much with cluster orientation relative to the laser polarization. The picture of the Mie plasmon resonance as an electron cloud vibrating against the positive ionic background suggests that the plasmon frequency should depend on the orientation in case of deformed clusters. Vibration along a wide extension should have a lower frequency (smaller restoring force) and vibration along a narrow side a higher one. This is indeed found in theoretical calculations and in experiment. As experimental results average over cluster orientations, one sees the effect as a significant splitting of the Mie plasmon. This collective splitting was already illustrated in Figure 2.7 and exploited for deformation analysis in Figure 4.5. We will address it here in more detail, as it is a practical way to access cluster shapes.

Let us briefly recall the reasoning from Section 2.3.3. In the case of spherical clusters, the Mie plasmon resonance is represented by one sharp peak. It splits into two peaks for axially symmetric deformed clusters. For an axially elongated (prolate) cluster, the oscillations along the elongation axis are weaker and thus have lower frequency than the two oscillation modes perpendicular to it. The situation is exactly reversed for an axially squeezed (oblate) cluster having two elongated axes and one short axis. This case is signaled by one peak at a higher frequency peak and two at lower frequencies. The latter is degenerated and the multiplicity is read off from the relative heights of the peaks. In the case of triaxial clusters, three peaks appear and they are inversely sorted to the length of the axes.

This collective splitting of the Mie plasmon was systematically used to measure deformations of small Na clusters as illustrated in Figure 4.8. Figure 4.8b shows

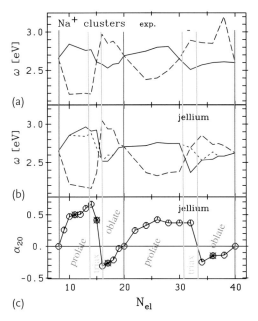

Figure 4.8 Deformation and collective splitting of the Mie plasmon resonance as functions of the number of electrons N_{el}, in Na_N^+ clusters ($N_{el} = N - 1$). Theoretical calculations are done with TDLDA (see Section 3.2.4) using soft jellium for the ionic background (see Section 3.3.2). (a) Frequencies of the two dominant peaks (major peak = solid, minor peak = dashed) from the experiments of [34, 99]. (b) Theoretical results for the frequencies of the Mie resonance peaks along x, y and z axis computed with the triaxial deformed jellium model. (c) Dimensionless quadrupole deformation α_{20} (3.14) for the ground-state configurations.

theoretical results, with an ionic jellium background (see Section 3.3.2), produced with the deformations as shown in Figure 4.8c. The three modes in spatial x, y and z directions are drawn, as clusters may be axial or triaxial in this size range. Two of these modes are degenerated for axially symmetric clusters (which are the majority), while three separate peaks indicate a triaxial shape. Figure 4.8a shows experimental measurements where the positions of the dominant peaks of the photoabsorption cross-sections have been deduced from a fit with Gaussian profiles. The solid line shows the strongest peaks which are to be associated with the theoretical peaks where two modes coincide. Axially symmetric and prolate clusters have the stronger mode (i.e., two modes) at higher frequency, while oblate clusters have it at the side of lower frequency. This is nicely confirmed by the figure. One can also note a very good agreement between theory and experiment, even if triaxial shapes are not yet clearly discriminated by the experimental data. This reflects, once again, the fact that it is extremely hard to experimentally disentangle the triple splitting in the rather broad absorption spectra.

This analysis through collective splitting of the resonance works reliably well for small clusters up to $N \approx 40$. Strong Landau fragmentation (as was discussed in Figure 4.7) sets on for larger clusters. This blurs the information about deformation splitting for medium large free clusters [35]. One has to be aware of that interference and use detailed modeling to disentangle collective splitting from Landau fragmentation. Extreme deformations and/or huge clusters still allow one to clearly separate the different branches of the Mie plasmon mode as, for example, for large gold nanorods or deposited clusters.

4.2.3
Thermal Shape Fluctuations

The yet missing mechanism is indicated in Figure 2.3 showing the optical response of Na_7^+ clusters at various temperatures. The spectrum at low temperature has three narrow and well-separated peaks. They emerge from a mix of collective splitting of the strongly oblate Na_7^+ with fragmentation by one close 1ph state. For larger temperatures, the peaks grow broader and the lower double peak merges into one. Thermal effects at the side of the electrons cannot induce such a width because 400 K = 0.03 eV is a small energy at the electronic scale. It is the thermal ionic motion which is responsible for this line broadening. Ionic vibration modes reside in the energy range of about 0.01 eV for Na_7^+. A thermal excitation of 400 K is thus a large amount on that scale. Therefore ionic motion is driven to rather large amplitudes and the cluster undergoes substantial thermal shape fluctuations. We have seen in Section 4.2.2 that deformation has a strong influence on the dipole spectrum. Thus each member of the thermal ensemble with its different deformation contributes a different spectrum to the total optical response. These spectra all incoherently add up to rather broad peaks.

4.2.4
The Width of the Mie Plasmon Resonance

So far, we have seen several mechanisms contributing to the width of the Mie plasmon resonance: Landau fragmentation, deformation splitting, and broadening through thermal shape fluctuations. We will discuss here two further broadening mechanisms and the systematics of Landau fragmentation.

The thermal broadening mechanism deals with incoherent superposition from spectra taken as snapshots at quasi-stationary samples of the ensemble. There exists in addition a coherent line broadening from dynamical coupling of the Mie plasmon to ionic oscillations (the "phonons" of the cluster). A coherent superposition of the plasmon with one or more phonon modes yields side bands to the plasmon peak with frequency shifts ≈ 0.01 eV, typical of ionic modes which had been experimentally observed in Li_4 [100]. Coupled electronic and ionic dynamics in TDLDA-MD (see Sections 3.1.3 and 3.2.4) automatically yields this coherent phonon coupling.

A further broadening mechanism is the effect of electron–electron correlations (coupling to 2ph and higher states). Similar to the case of electron–phonon coupling, one obtains a further fragmentation of the TDLDA spectrum (see Section 3.4.4) over a small neighborhood of less than 0.1 eV. The computation of coherent 2ph correlations is extremely cumbersome and requires highly developed many-body methods. For example, CI calculations include such higher correlations and they indeed show some broadening effects [64], but their strength is small and does not contribute much to the width of the plasmon. However, dynamical electron–electron correlations are known to become especially important for strong excitations beyond the linear regime. The Uehling–Uhlenbeck collision term has been designed to account for that effect, see Section 3.2.5.

After all, Landau fragmentation and incoherent thermal line broadening remain the most important mechanisms creating the width of the Mie plasmon peak. Thermal broadening shows, of course, a strong temperature dependence, while Landau fragmentation does not. The latter, however, has an interesting dependence on system size. This is demonstrated in Figure 4.9 showing the trend of the full width at half maximum (FWHM) of the surface plasmon resonance with varying system size. Only results from spherical clusters are drawn such that we basically see the effect of Landau fragmentation. The TDLDA width is practically zero for small clusters ($N \ll 1000$), increases to a maximum around 1000, and then decreases linearly with $N^{-1/3}$. The increase of the FWHM is due to the increase of the density of 1ph states with growing cluster size. The decrease for $N > 1000$ is surprising as it develops in spite of increasing density of 1ph states. What happens is that the coupling matrix elements of the collective resonance with the 1ph states shrink towards bulk material. In fact, for a bulk plasma, it is known that the plasmon strictly decouples from the particle excitations [8]. The trend linear in $N^{-1/3}$ in the regime $N > 1000$ can actually be derived analytically in a model of plasmon waves scattered by the cluster surface. Figure 4.9 also shows experimental results. The trends

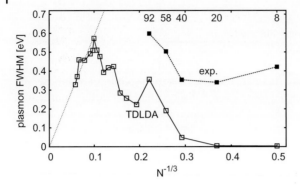

Figure 4.9 Trends of surface plasmon widths (FWHM) for Na_N^+ versus $N^{-1/3}$ (which is proportional to the inverse of the system radius). Compared are results from TDLDA calculations (solid line, open boxes) [98] with experimental data (dashed line, filled boxes) from [97]. The fine dashed line indicates a linear fit to the asymptotic $\propto N^{-1/3}$ for large N which corresponds to electron scattering from the walls of the Kohn–Sham potential [101].

are qualitatively reproduced by calculations. But there is a large offset for small clusters which is probably due to thermal broadening.

4.3
Photoinduced Electron Emission

Irradiation of a cluster by an intense laser pulse immediately leads to electronic emission, which, as already discussed in Section 3.4.5, is basically explored at three (connected) levels: total net ionization, photoelectron spectra (PES) and photoangular distributions (PAD). We illustrate each of these observables in a few examples in the following.

4.3.1
Total Ionization

Net ionization following laser excitation is the simplest signal from electronic emission. Even so, it carries a great deal of useful information. For example, ionization as a function of laser frequency for fixed intensity maps, to a good approximation, the frequency dependence of the photoabsorption cross-section, as long as small or moderate laser intensities are used. Ionization as a function of laser intensity I for fixed frequency follows a trend I^ν, where ν is the effective number of photons involved in the process. The number ν is naively the next higher integer to $E_{IP}/(\hbar\omega_{las})$, where E_{IP} is the energy of the ionization potential. This simple estimate, however, is often overruled by reaction pathways proceeding through resonant excitation, thus lowering the effective photon number. This I^ν law applies in the regime of moderate intensities and sufficiently short pulses. The trend levels off in the strong field regime. Electronic thermalization becomes increasingly im-

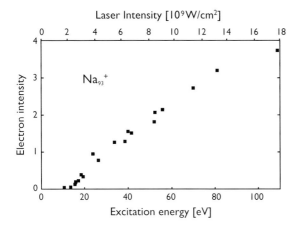

Figure 4.10 Experimental electron intensity (that is, ionization) versus excitation energy (equivalent to laser intensity, see upper horizontal scale) for laser excitation of Na$_{93}{}^+$ in the regime of thermal emission. Adapted from [102].

portant for longer laser pulses (above 20–100 fs depending on the material). In that case, nearly all the absorbed photon energy is first converted into intrinsic heat, and released much later by thermal electron emission, or by monomer (neutral atom) evaporation as a competing channel. Figure 4.10 presents an example of thermal emission, the trend of ionization versus laser intensity for Na$_{93}{}^+$ excited with ns pulses. This curve shows a monotonic increase of ionization with excitation energy. The data are well reproduced by the simple Weißkopf rule, see Eq. (3.18), (not shown here) deduced from the assumption of full thermalization of the excitation energy and subsequent thermal emission.

Total ionization is a key signal in more involved excitation setups. A prominent example is given by pump-probe experiments. Typically, a pump pulse excites the molecule or the cluster, thus triggering ionic motion (oscillation or fragmentation). The changing ionic background produces sensitive changes of the electronic spectrum. A probe pulse explores these electronic changes by recording the ionization yield as a function of delay time. An example will be given in Figure 5.6.

4.3.2
Photoelectron Spectra (PES)

The basic mechanisms of PES have already been illustrated in Figure 2.9 and discussed in Section 2.3.4. For completeness, we recall here the basics of PES. The situation is very simple in the regime of one-photon processes: the photon energy $\hbar\omega_{\text{las}}$ is added to the single-particle (s.p.) energy ε_i of an occupied state, thus mapping the s.p. spectrum to kinetic energies visible as peaks in the PES, following $\varepsilon_{\text{kin}} = \varepsilon_i + \hbar\omega_{\text{las}}$, which is directly visible in Figure 2.9. This is a widely used tool of analysis in atomic, molecular, and cluster physics. For example, the density of states shown in Figure 1.6 was produced this way. This "simple" picture basically gives information on structural properties of clusters. But PES are much richer than that, since it can readily address dynamical situations. We illustrate this aspect on a few examples below.

One-photon emission first requires that the photon energy is larger than the IP, and ideally larger than the whole span of single-particle energies to be mapped. For lower photon frequencies, direct electron emission is possible only if more photons cooperate. This leads to the regime of multi-photon ionization (MPI) (see also the discussion in Section 4.4.2 on laser ionization mechanisms). Figure 4.11 shows an example of PES in the MPI regime in the simple test case of Na_9^+ in which the s.p. spectrum is sufficiently simple to allow a clear picture of MPI effects. The occupied states grouped into two bunches: a one-electron ground state structurally close to a 1s state in a spherical potential well, and a couple of nearly degenerated states which we label as 1p. We have extended to negative values the energy axis in the plot thus allowing to indicate these two s.p. states. We also indicate how a couple of photons are stacked on top of each other to move the electrons into the continuum. This produces a series of copies of the s.p. spectrum in the PES peaks appearing at $\varepsilon_{kin} = \varepsilon_i + \nu\hbar\omega_{las}$ where ν is the photon number, see Eq. (2.7). The case of the moderate intensity $I = 10^{10}$ W/cm^2 (light gray in Figure 4.11b) shows these peaks for $\nu = 3, 4, 5,$ and 6. Processes become more unlikely the more photons are involved. Thus the yield quickly decreases with increasing ε_{kin}. Figure 4.11 indicates the envelope of the PES signal by a dashed line and this line exponentially decreases, approximately following $\exp[\nu \log(I/I_0)]$, where I_0 is some parameter and $\log(I/I_0) < 0$.

Figure 4.11b also shows a result for a larger intensity, $I = 10^{11}$ W/cm^2. The peak structure is then wiped out. The reason is that this case comes along with stronger ionization which enhances the Coulomb attraction and thus deepens the attracting field at the cluster site. The mechanism is sketched in the upper left corner of Figure 4.11a. It indicates how the occupied states move gradually down in energy due to ionization. The final PES is a superposition of emission from all stages along the ionization process. Averaging the peaks with this shift width leads to a smooth pattern as observed in the figure. The exponential decrease remains, however, with a smaller slope. Indeed it is obvious from the trend $\exp[\nu \log(I/I_0)]$ that higher intensity I renders $\log(I/I_0)$ less negative, and so reduces the slope. In fact, the pattern now looks purely exponential. This could easily be misinterpreted as thermal emission. But the example shows that exponential PES alone does not suffice to distinguish directly from thermal emission. Only PAD can help to decide because thermal emission is strictly isotropic while direct emission usually has a pronounced anisotropy.

It is interesting to relate that to the total ionization yield. As a rule of thumb, a total ionization of about 1 defines the transition region between the frequency-dominated perturbative regime below, and the field-dominated nonlinear regime above.

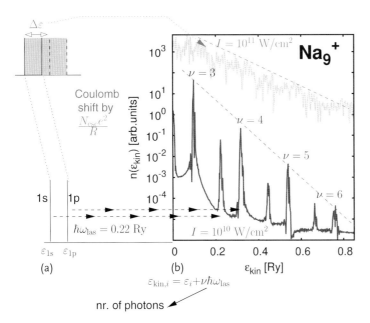

Figure 4.11 Schematic representation of PES in the regime of multiphoton ionization for the Na$_9^+$ cluster excited by a laser with frequency ω_{las} of 3 eV (= 0.22 Ry), pulse width of 48 fs (a), and two different intensities (b) of $I = 10^{10}$ W/cm^2 (dark gray line) and $I = 10^{11}$ W/cm^2 (light gray line). Comments drawn into the figure are explained in the text.

4.3.3
Photoelectron Angular Distributions (PAD)

4.3.3.1 PAD in the Perturbative Regime

As long as the laser parameters remain in a regime of linear or weakly nonlinear excitations, ionization processes can still be described by perturbation theory [103]. The PAD of electrons emitted from a (spherical) electronic system can in general be expanded as:

$$\frac{d\sigma}{d\Omega} = \frac{\sigma}{4\pi}\left[1 + \beta_2 P_2(\cos\vartheta) + \beta_4 P_4(\cos\vartheta) + \cdots + \beta_l P_l(\cos\vartheta)\right], \quad (4.1)$$

where P_l are Legendre polynomials and β_l the anisotropy parameters. The distribution is axially symmetric (depending only on the angle ϑ) because the laser pulse together with a spherical system represents an axially symmetric situation. The maximum possible value of $l = \nu$ corresponds to the order of the photon process needed to reach ionization, see Eq. (2.7). For one-photon processes, the distribution thus reduces to

$$\frac{d\sigma}{d\Omega} = \frac{\sigma}{4\pi}(1 + \beta_2 P_2).$$

The value of β_2 ranges in that case in the interval $-1 \le \beta_2 \le 2$. The case $\beta_2 = 2$ corresponds to a $\cos^2\vartheta$ shape, which means an electron emission perfectly aligned

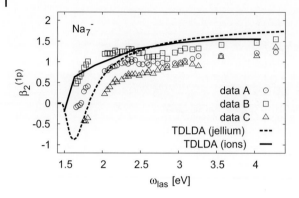

Figure 4.12 Anisotropy parameter β_2 of the single-electron state 1p, as a function of laser frequency ω_{las}, in $Na_7{}^-$. Comparison between TDLDA with ionic background or jellium background (see Sections 3.2.4 and 3.4.4), and experiment. The three groups of data correspond to the three slightly different sublevels of the 1p state. Adapted from [104].

with the laser polarization. The case $\beta_2^{(i)} = -1$ corresponds to a $\sin^2 \vartheta$ shape, arising from an electron emission perpendicular to the laser polarization.

The term "spherical system" has yet to be clarified. Atoms are spherical but molecules, and even more so clusters, are not. However, most measurements deal with an ensemble of free, randomly oriented molecules or clusters. Such an ensemble is effectively isotropic and the form of PAD is again given as in (4.1). A theoretical description of such ensembles first deals with the deformed cluster in detail and one has then to perform an orientation averaging.

Interestingly enough, the characteristics of the PAD for one-photon processes can thus be packed into one single parameter, namely the anisotropy parameter β_2. Being able to sum up all information into only one parameter allows to display trends in a simple fashion. The point is illustrated in Figure 4.12 where the anisotropy parameter of the 1p state in $Na_7{}^-$ is plotted as a function of laser frequency. The figure displays three sets of experimental results and two theoretical estimates with different modeling of the ionic background. The striking feature of Figure 4.12 is the pronounced dip in β_2 close to ionization threshold at 1.43 eV, equally seen in data and both models. This is interpreted as being due to the faint binding of the anion where the outgoing electron cloud is practically a free plane wave. It is also worth noting that the theoretical approaches globally agree with the experiment but differ in detail. This indicates that the results depend on ingredients of the modeling, and reliable predictions of PAD require a high level of description.

4.3.3.2 Mixed PES and PAD from Velocity Map Imaging

As already briefly mentioned at the end of Section 2.3.4, velocity map imaging (VMI) allows to depict in a compact manner combined PES/PAD through a polar representation: the photoelectron kinetic energy increases along the radial direction, and the emission angle ϑ is measured with respect to the laser polarization.

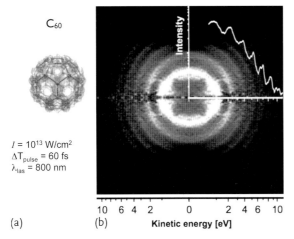

Figure 4.13 The structure of C_{60} is shown (a), as well as a velocity map image in the case of C_{60} (b) irradiated by a laser with parameters as indicated, in a polar representation with radii corresponding to photoelectron kinetic energies and polar angle to the emission angle with respect to the laser polarization (horizontal direction). The photoelectron yield is represented as a shaded plot in logarithmic scale where dark stands for little emission and light for high yield. The inset illustrates an angle-integrated PAD, that is the PES. Adapted from [84].

We here discuss in more detail an example of a VMI from irradiated C_{60}, shown in Figure 4.13. The figure is rich on information. One particularly interesting aspect is the evolution of PAD with number of photons ν. Each ring in the plot corresponds to a given ν: the higher it is, the larger the radius. Note that the corresponding PAD are nearly isotropic for low ν and become more forward/backward directed, the more photons are involved. The effect can be understood in two ways. In the transient regime of moderate laser intensities, we can argue with multiphoton perturbation theory. A strong focus on small ϑ requires high $P_l(\cos\vartheta)$ components in the PAD. Each further photon enhances l by one and thus enhances contributions from high photon numbers ν required to produce more forward/backward dominance. The argument is simpler in the strong field regime. The external field simply pulls the electrons in the field direction, the stronger the field.

4.4
Cluster Nonlinear Dynamics

The regime of linear response is rather predictable because practically all information is contained in the spectrum of excitation modes. Nonlinear processes unfold an extremely rich world of scenarios. A comprehensive discussion exceeds the bounds of this book. We will here briefly sketch a few highlights of the field. Nonlinear excitations can be reached in several ways, using either photons or particles, typically ions. The by far most developed tool of excitation are lasers, which exhibit an enormous versatility and require rather modest experimental setups, at least for standard intensities. We shall thus focus the discussion in this section on the case of laser irradiation on clusters.

4.4.1
Tunability of Lasers

Lasers, or photons in general, represent an extremely versatile tool for excitation. Already the simple pulse (2.2a)–(2.2c) has three different parameters to choose: intensity, pulse duration, and frequency. Even more flexibility comes into play with dual pulses (used, for example, in pump-probe experiments) or dedicated chirp (small variation of frequency during the pulse). The enormous progress in the development of light sources these days provides a very broad choice of photon parameters. Figure 4.14 schematically illustrates the various regimes and corresponding machines in the plane of laser frequency and intensity. One sees the enormous large intensity range of optical lasers. The range of available conditions is further extended dramatically by free-electron lasers (FEL), which exist for photons in the infrared, VUV and X-ray regime.

These photon properties now have to be related to the electronic reaction. The impact of the photon can be conveniently characterized by the ponderomotive potential U_p. Consider a freely oscillating electron (pure quiver motion, no drift velocity) in a laser field $E_0 \exp(i\mathbf{k} \cdot \mathbf{r} - i\omega_{\text{las}} t) f(t)$, see Section 2.2.2. Its ponderomotive potential U_p represents the electron kinetic energy averaged over a photon cycle. At

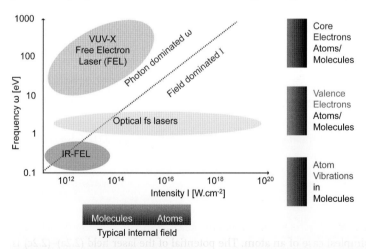

Figure 4.14 Schematic representation of various dynamical regimes as a function of laser intensity I and photon frequency ω. The dashed diagonal line represents frequency-intensity combinations with constant Keldysh parameter $\gamma = 1$, see text for details. This line characterizes the transition from photon-dominated to field-dominated regime for an assumed IP of a few eV. The blocks to the right side indicate typical frequency ranges as labeled. The block below the plot indicates typical atomic field strengths related to given laser intensities.

the maximum of the laser envelope ($f = 1$), it reads as

$$U_p = \frac{e^2 E_0^2}{4 m_{el} \omega_{las}^2} = 9.33 \times 10^{-14} \text{ eV} \times I_0 \, [\text{W/cm}^2] \, (\lambda[\mu m])^2 \quad (4.2)$$

where $E_0 = |\mathbf{E}_0|$ and λ is the photon wavelength. The other aspect in the reaction is the electronic binding in the cluster. This can be quantified by the ionization potential (IP) and the associated energy (E_{IP}), which corresponds to the energy cost for removing one electron. What counts is the relation between the E_{IP} and U_p, which is expressed in the Keldysh parameter:

$$\gamma = \sqrt{\frac{E_{IP}}{2 U_p}} = \sqrt{\frac{2 E_{IP} \omega_{las}^2}{I_0}} \, . \quad (4.3)$$

The case $\gamma \ll 1$ stands for the field-dominated regime where direct ionization (over barrier or tunneling) prevails, while $\gamma \gg 1$ signals the frequency-dominated regime of weak perturbations where emission proceeds through multiphoton ionization. The dividing line $\gamma \approx 1$ is indicated in Figure 4.14 assuming an IP of $E_{IP} = 1$ eV. Typical frequency ranges of atomic and molecular systems are indicated to the right side of Figure 4.14. The lowest frequencies in the deep infrared are associated with molecular vibrations. The range around visible light belongs to the dynamics of valence electrons. Core electrons move at much higher frequencies in the X-ray regime. The intensity is proportional to the squared field strength. The gray box below the plot indicates typical atomic and molecular field strengths in terms of an equivalent intensity. This demonstrates that lasers with moderately high intensity can ionize a molecule at once by simply overriding the binding forces (field-dominated regime).

In addition to what Figure 4.14 shows, these light sources have time profiles which can be tuned in a certain range. Most flexible are optical lasers with pulse lengths from nanoseconds down to attoseconds [105]. They also allow multiple pulses with definite time delay and dedicated chirps (see Section 4.4.4). FEL are yet on their way to comparable flexibility. Presently, pulse lengths down to 20 fs are feasible. It is likely that even shorter pulses will be developed in the future.

4.4.2
On Ionization Mechanisms

Basic mechanisms for laser ionization are sketched in Figure 4.15. Figure 4.15a shows the simplest case of an atom. The potential of the laser field (2.2a)–(2.2c) is linear over the dimensions of the atom. It generates a barrier as indicated in the plot. If the laser force field is comparable to the atomic binding forces, the barrier becomes so low that tunneling through a small barrier becomes likely. For larger forces, even immediate electron emission over the barrier takes place. This is the regime of optical field ionization. A different mechanism is at work for lower field strengths. An electron absorbs energy from several photons at once until it has

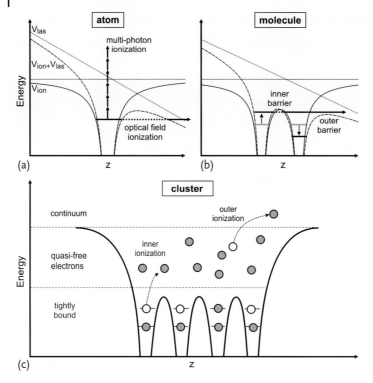

Figure 4.15 Schematic view of ionization mechanisms. In all panels occupied electron states are indicated by horizontal bars. The first one demonstrates the pathways of multiphoton ionization optical field ionization on the example of an atom (a). Next, the distinction between an inner ionization barrier and an outer barrier on the example of a dimer molecule is illustrated (b). In both (a) and (b), the potentials of the unperturbed ions V_{ion} (solid line), of the laser V_{las} (dots), and of the sum of both (dashes) are drawn. Also represented are the effect of inner and outer ionization in the case of a cluster (c). The effective electron potential (without laser) is shown as a solid line. From [84].

gathered a sufficient amount to reach the continuum. This process is called multiphoton ionization (MPI). Traces of MPI can be seen in photoelectron spectra as, for example, in Figure 4.11 where the emission processes with different numbers of photons can be clearly resolved. As discussed above, a useful measure for the actual regime is the Keldysh parameter (4.3), where direct field ionization dominates for $\gamma \lesssim 1$ and photon-induced ionization for $\gamma \gg 1$.

The situation becomes more involved for composite systems. The simplest example here is a dimer molecule as shown in Figure 4.15b. Besides the competition of laser field and overall binding, here we additionally have the fields exerted by the neighboring atom. This can give rise, under appropriate geometrical conditions, that is, an optimal distance between atoms, to enhanced ionization, a process called charge resonance-enhanced ionization (CREI) [106]. Furthermore, we have now to distinguish between inner ionization and outer ionization. Inner ionization releases a tightly bound electron from the atom to freely travel through the molecule. Outer ionization describes the full separation of the electron from the whole molecule.

This distinction becomes more important in large molecules and clusters, as indicated in Figure 4.15c. The electrons stemming from inner ionization then contribute to the long-range oscillations of the valence electron and so, amplify collective effects as, for example, plasmons. At moderate laser intensities, systems with initially delocalized electrons, as metal clusters, may undergo outer ionization only. Inner ionization increasingly contributes with increasing laser intensity. This gradually wipes out the initial difference between the materials and, for example, renders rare gas clusters as reactive as metal clusters. The onset and self–amplification of these additional ionization processes as a result of previously created quasi-free electrons, and cluster outer ionization has been coined ionization ignition.

4.4.3
Production of Energetic Ions and High Charge States

Irradiation of clusters by intense laser beams has attracted much attention in the past. It turns out that clusters have an enormous capability to store energy from the laser (much more than the corresponding gas of atoms), to be related to the different ionization mechanisms just discussed. As a result of this energy storage, energetic particles, electrons, photons, ions, are emitted from such a process. This usually leads to accumulation of high charge states and thus a final Coulomb explosion of the system. We have already discussed the case of high energy photons (X-rays) in Section 2.3.3. Here, we present the example of ion production. Due to strong charging on a femtosecond time scale, substantial Coulomb energy is accumulated, leading to high recoil energies in the subsequent explosion of the system. In one of the most exciting examples, these fast ions have been exploited to trigger fusion [107].

Clusters which are excited by intense fs pulses emit ions with surprisingly high kinetic energies. As an example, Figure 4.16 shows recoil energy spectra obtained from laser excitation of Ar_N for different cluster sizes N [108]. The distributions show a plateau with kinetic energies extending between a few keV up to 30 keV, followed by a steady decrease for higher energies. It is interesting to note that the tail is shifted to higher energies with increasing cluster size. This indicates that a collective cluster effect is responsible for these high energies.

To give some insight of the underlying mechanisms, we consider the cluster as a homogeneous matter distribution. The intense laser pulse leads to strong ionization in a short time span. An ion thus sees the Coulomb field of a homogeneously charged sphere as an initial state for its escape. Its final kinetic energy ϵ_I is determined by its initial potential energy:

$$\epsilon_I(r) = \frac{4\pi}{3} n_I R_I^2 q^2 \times 14.4 \text{ eV Å} , \tag{4.4}$$

where R_I is the initial radial position of the ion, q the charge state of the cluster, and n_I the number density of ions in the cluster. Therefore, ions located at the

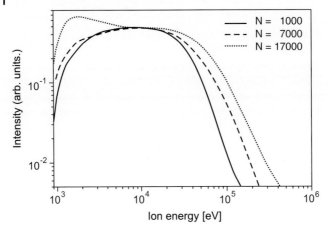

Figure 4.16 Ion recoil energy spectra from Ar$_N$ with average cluster sizes $N = 1000$ (solid), 7000 (dashed), and 17 000 (dotted), for exposure to laser pulses with intensity of 1.3×10^{16} W/cm^2, duration of 50 fs and wavelength of 790 nm. After [108].

cluster surface acquire the highest recoil energies $\epsilon_I^{\max}$ and this maximum value naturally increases with cluster size. These trends have been observed in several experiments of laser-driven Coulomb explosion of clusters.

More information on emitted ions is shown in Figure 4.17 for the case of a highly excited Pb cluster. The spectrum of kinetic energies has been recorded for each charge state separately. The figure shows the average kinetic energies versus charge state. Two regimes can be recognized. For low charge states q, the kinetic energy quadratically depends on q. This complies with the above explained model of Coulomb explosion of an initially localized ensemble of Pb ions with charge q. The trend turns to a linear dependence on q for high charge states. This is interpreted in [109] as coming from hydrodynamical flow of the highly charged ions following the expansion of the nanoplasma.

The example indicates that the situation is not always as simple as suggested by the above schematic considerations. The processes in the build-up of the initial state depend on the material. The starting radius for the expansion can also sensitively depend on the laser parameters. As already discussed at the end of Section 4.4.2, after ionization ignition, the differences between materials becomes unimportant. While initially metals have free electrons and rare gases do not, here we always encounter an optically active gas of free electrons. In the next stage, the actual laser profile determines the evolution. The Coulomb pressure generated by ignition leads to a steady expansion of the system, slower or faster depending on the charge state after ignition. The frequency of the plasmon associated with the electron gas decreases with increasing system radius until it comes into resonance with the laser frequency. At this point, the laser-cluster interaction becomes very strong. It then triggers a very fast and violent ionization which produces the initial stage for the Coulomb explosion sketched above.

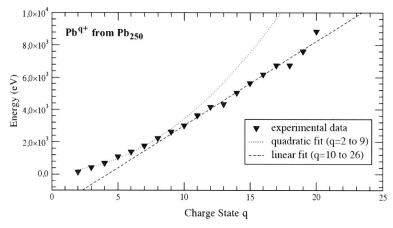

Figure 4.17 Average kinetic energies of Pb^{q+} ions emitted from an exploding Pb cluster versus charge state q. Results are an average over 1000 events with laser pulses with full width at half maximum of 760 fs, intensity of 3×10^{15} W/cm^2, and wavelength of 790 nm. The average cluster size is 250. The dashed line shows a linear trend fitted to high q data and the dotted line a quadratic trend applicable for low q data. After [109].

Various models have been developed to describe the cluster dynamics in such a violent regime. A fully quantum mechanical treatment as, for example, TDLDA-MD is far beyond feasibility. The complexity of the problem calls for simple schemes. One typical (and probably the most prominent) example for modeling in the regime of violent laser excitations of large clusters is the nanoplasma model. It starts with a macroscopic hydrodynamical picture for the flow of ions and electrons augmented with rate equations for transitions between species (atoms, ions of different charge). The relevant dynamical variables then reduce to a set of integral quantities (radius, ionization, etc.). Such an oversimplified picture aims at describing just the gross properties of the system. This is appropriate for the case of high intensity irradiation of clusters because the details of the excitation mechanism tend to be washed out to the benefit of a global plasma-like behavior, the nanoplasma. Properly calibrated, the model can cover the complete scenario of the interaction processes, taking into account ignition, ionization, heat flow, polarization fields, electronic emission and cluster expansion. For a more detailed summary, see Section 5.2.2 of [47].

4.4.4
Variation of Pulse Profile

The above discussion of violent cluster excitation has indicated an intimate balance between ionization, system expansion and resonance conditions. This balance can be varied systematically in pump-probe experiments, which are widely used in exploring the ionic dynamics of molecules [26]. While such a time-resolved analysis is often tracking detailed reaction pathways, the complexity of clusters mainly allows

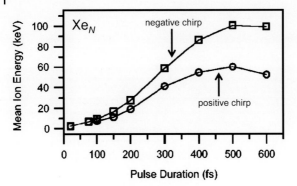

Figure 4.18 Mean recoil energy of atomic ions emitted from Xe_N clusters of average size $\langle N \rangle \approx 5.5 \times 10^4$ for excitation with stretched pulses at 800 nm and constant peak intensity of 2×10^{17} W/cm^2. The data shows results obtained with positively and negatively chirped pulses (see text for details). Adapted from [111].

to follow a few global observables as overall radius or deformation (see Figure 5.6 in Section 5.1.2).

The pulse length of a single pulse can already influence the yield in reactions after strong laser excitation [110]. In fact, all laser parameters are at one's disposal for achieving optimal control over the pursuit of a reaction. A very promising feature in that respect is the chirp (i.e., time-dependent frequency) of a laser pulse.

Figure 4.18 demonstrates the effect of two pulse parameters, namely pulse length and chirp, on an experiment with large Xe_N clusters. The mean ion energy, indicating the violence of the reaction, first grows with pulse length up to an optimal value around 500 fs. It then turns to decrease again. This is due to a resonance mechanism similar to that discussed in the previous section for the Coulomb explosion of Pb clusters. The laser starts with gently ionizing the clusters which triggers a slow expansion. A violent ionization begins if resonant conditions are reached. This requires that the pulse lasts long enough to be still active if resonance occurs. If the pulse becomes too long, resonance conditions fade away again.

Figure 4.18 also shows the effect of a chirp within the pulse. For negative chirp (laser frequency decreasing over the pulse), a 60% enhancement of the mean ion energy was observed. Negative chirp means that it takes longer for the expanding cluster to reach resonance conditions. This is indeed seen from the shift of the maximum to longer pulses. This, in turn, provides a longer time span for resonant collective absorption: the joint gradual frequency red-shift of both laser pulse and resonance consequently produces a larger yield.

In spite of the great complexity of cluster dynamics, there also exist a few successful examples of true pump-probe experiments in clusters. We will discuss an example in connection with Figure 5.6. This was an experiment where the Ag cluster was kept inside a He droplet for reasons of better handling. Therefore, this case is found in Section 5.1.2.

5
Clusters in Contact with Other Materials

This chapter proceeds towards applications of clusters in nanoscience and technology. Most practical applications deal with clusters in contact with some environment. Thus, we will first address the topic of clusters embedded in, or deposited on, a substrate, explain their theoretical description and discuss a few basic properties. After that, we present selected examples of applications of clusters in nanoscience.

5.1
Embedded and/or Deposited Clusters

5.1.1
On the Relevance of Embedded or Deposited Clusters

So far, we have discussed free clusters in order to understand the cluster properties as such. However, free clusters represent only a fraction of possible systems involving clusters. A rich world of effects emerges from combination of materials, here in particular the combination of clusters with substrates, either deposited or embedded. On the other hand, that combination will complicate the theoretical description. Thus, there are several good reasons to specifically address the complex problem of such combined systems. A great manifold of experiments is actually performed for clusters in contact with a substrate, either deposited on or embedded in, see for example [112]. A major reason is that many experiments can better be performed for nonisolated systems. Indeed, substrates serve to prepare well-defined conditions of temperature and orientation, they help to handle neutral clusters, and they allow to gather higher cluster densities.

A further important aspect is that clusters in contact with an environment is a realistic scenario in many applications. For example, there are promising attempts to employ clusters in the dedicated shaping of nanoscaled devices [113]. Small Au clusters on surfaces are also found to be efficient catalysts [114]. Moreover metal clusters are considered as nanojunctions in electrical circuitry [115]. And finally the coupling to light is exploited in producing an enhanced photocurrent by depositing Au clusters on a semiconductor surface [116], to cite only a few prominent examples.

An Introduction to Cluster Science, First Edition. Phuong Mai Dinh, Paul-Gerhard Reinhard, and Eric Suraud.
© 2014 WILEY-VCH Verlag GmbH & Co. KGaA. Published 2014 by WILEY-VCH Verlag GmbH & Co. KGaA.

Metal clusters in inert substrates are also simple model systems for chromophores where the field amplification effect has a large impact on the environment, see for example the study of localized melting for the generic combination of Au clusters embedded in ice [117]. Such combinations can be used as a test system to understand the first stages of radiation damage starting with defect formation in solids [118]. But there are also promising applications in medicine where the frequency selective optical coupling of organically coated metal clusters attached to biological tissue may be used as "photosensitizers" for imaging, diagnosis and for localized heating in cancer therapy, see Section 6.4.

Last but not least, compounds of two different materials are a research topic in its own right as the mutual influence of the two species can create new effects which were not possible for the isolated species. We will sketch such effects at many places below (see for example Section 5.1.2). A salient example are biological applications where the impact of the ever present water environment is known to be crucial (see Figure 5.4). It is obvious that the interactions between the cluster and its environment are crucial and that a proper description requires a careful treatment of these interactions. Clusters in/on substrate are, of course, much more complex systems than free clusters and thus constitute a great challenge for theoretical modeling. The large number of atoms in a substrate inhibits its fully quantum mechanical description. In order to preserve most quantum effects (at least where/when they are mandatory), one usually relies on a mixed quantum mechanical/molecular mechanical (QM/MM) modeling. In QM/MM approaches, small dynamically active regions are computed fully quantum mechanically, while farther away and/or more inert regions employ simple molecular mechanics. This method was first developed in order to deal with the very complex and demanding systems of biochemistry [119] and is often used in that context. It is also applied very much in surface chemistry [120]. In any case, QM/MM approaches are extremely useful because they dramatically simplify the simulation of complex systems, or make it actually feasible. A central issue in many of those systems is to define a proper distinction between the QM and the MM parts. The case of clusters deposited/embedded on inert material is particularly simple in that respect, as it carries a natural distinction (cluster = QM, substrate = MM). The merits of applying a QM/MM approach to cluster physics thus are twofold. On the one hand, it allows the description of large samples of the substrate, rendering the simulation more realistic. On the other hand, clusters provide an ideal testing ground for QM/MM due to the better defined separation and more controllable experiments. The essence of QM/MM methods will be addressed in Section 5.2.1. We will then present in more detail the simple example of metal clusters in contact with an inert environment in Section 5.2.2.

5.1.2
The Impact of Contact with Another Material

Figures 5.1–5.7 provide examples of studies involving clusters in contact with an environment. They illustrate how the response of the cluster is affected by the presence of the environment and, in turn, how the cluster modifies the environment itself. We consider here various dynamical scenarios involving both linear and non-linear responses.

5.1.2.1 Examples in the Linear Regime

We start with Figure 5.1 with an example from standard optical absorption or "color" (for free clusters, see Sections 3.4.4 and 4.2), exploring the impact of the environment on the cluster linear response. The system is a small metal cluster Ag_8 embedded in a finite size rare gas matrix [121]. The evolution of the peak position with matrix size is plotted. The response of the Ag_8 cluster is indeed affected by its surroundings. But the effect is small: the ordinate scale is in nm, which corresponds to a total variation of order of 0.01 eV, while the surface plasmon resonance has an energy of a few eV. This presumably reflects the fact that the "matrix" is composed of rather inert rare gas atoms. Still, one can qualitatively spot different trends (increase/decrease) when considering different rare gases, an effect which is closely related to the increase of rare gas polarizability with increasing element number (from Ne to Xe).

Figure 5.2 illustrates a complementing aspect of the interactions between a cluster and its environment. Rather than analyzing the impact of the environment on the cluster, the example considers the modifications of a substrate by the presence of clusters deposited on it. The test cases are Pd clusters deposited on an insulating MgO surface. The observables are the energies of electronic levels of MgO, which are recorded as a function of Pd coverage [122]. As is clear from the figure, the MgO levels are significantly affected by the deposition process, in spite of the fact that MgO is a well-bounded insulator. This again points out that the subtle interaction effects can go both directions.

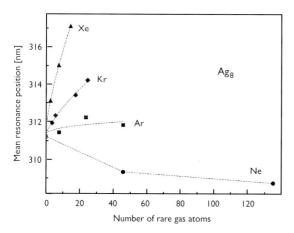

Figure 5.1 Optical response of a Ag_8 cluster coated by a finite number of rare gas (Ne, Ar, Kr, and Xe) atoms: position of the surface plasmon resonance as a function of the number of coating rare gas atoms. After [121].

104 | *5 Clusters in Contact with Other Materials*

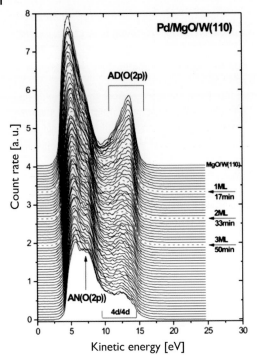

Figure 5.2 Impact of deposition of Pd clusters on electronic properties of the MgO substrate. See text for details. From [122].

5.1.2.2 Examples in the Nonlinear Regime

We now switch to dynamical scenarios exploring a nonlinear response. Figure 5.3 illustrates dynamical effects on embedded clusters and their environment. It shows the optical response of embedded clusters (Ag clusters inside bulk glass matrix) before (solid lines) and after (dashed and dotted lines) a violent laser irradiation [123]. The Ag clusters are grown by diffusing Ag^+ ions into glass, leading to a formation of Ag clusters in the cavities of the glass matrix. These metal clusters are rather compact and spherical. They thus exhibit a single resonance peak at high energy (short wavelength), which is degenerated in all laser directions, see solid line in Figure 5.3. The system is then exposed to a strong laser pulse (intensity of about 10^9 W/cm^2, duration of 150 fs). This changes the optical response spectrum dramatically. The peak position shifts to lower energy, which indicates that the cluster radius has increased (see Section 2.2.3). Moreover, the spectrum is split into two peaks, demonstrating that the clusters are deformed (see Section 4.2.2). The violent laser pulse has thus expanded and elongated the cluster. A free cluster would recover its original shape after the pulse is over. But the glass environment manages to freeze the cluster in the new shape produced by the laser pulse, as indicated by the modified optical response.

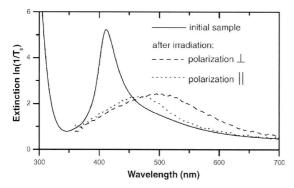

Figure 5.3 Optical response of a Ag cluster embedded in a bulk glass matrix, before (solid line) and after a violent laser excitation with two different polarizations (dashed and dotted lines). From [123].

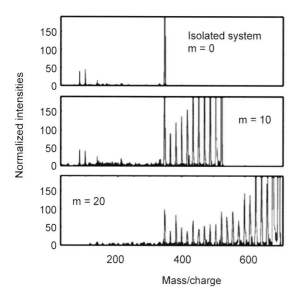

Figure 5.4 Fragmentation patterns of a biological molecule (here the adenosine monophosphate nucleotide) coated with a finite number m of water molecules, induced by collisions with energetic neutral atoms. Adapted from [124].

We explore in Figure 5.4 an even more violent scenario. The test case is a biomolecule coated with a finite number of water molecules [124]. Collisions with neutral atoms provoke the fragmentation of the complex. The fragmentation spectrum, plotted as a function of mass over charge ratio, exhibits a sizable dependence on the number m of coating molecules. The example demonstrates the intricate relation between system and "matrix" in this example of biomolecular systems, even in the course of violent disintegration.

5.1.2.3 Emergent Magnetism in Deposited Nanostructures

Magnetism is an aspect of cluster science which we cannot discuss in detail in this introductory book. But we indicate its great importance by one exciting example. A key question is the onset of magnetic ordering in nanostructures of materials that are nonmagnetic in bulk. The point can be generalized to any material that is not a traditional d-band magnet and is known as emergent magnetism. Unexpected magnetism has been reported for a variety of materials, including carbon, graphene, organic monolayers, silicon, nanocrystals of metals like Au and metal oxides. The emergence of magnetism depends on the materials, but is generally linked to defects, doping, surface effects and quantum size effects [125].

We focus here on the role of substrate effects on the emergence of magnetism in Rh nanostructures. While Rh atoms have a finite magnetic moment, bulk Rh does not. Moreover, experimental and theoretical investigations have shown that free (gas phase) Rh clusters do exhibit a magnetic moment for cluster sizes below 100 atoms per cluster. Recent experiments suggest that magnetization is localized on the cluster surface and that magnetism disappears, once Rh clusters are deposited on a Ag or Pt surface [125]. These results reflect a critical dependence of the magnetism in Rh clusters on cluster structure and the surface orientation of the substrate. However, while Rh clusters in direct contact do not exhibit magnetism, a net magnetization appears if the cluster growth is decoupled from the substrate (Ag, Pt) by a Xe buffer layer. The point is illustrated in Figure 5.5 where we show the total moment per Rh atom for the Rh clusters on a Xe buffer on top of an Ag surface. One deduces magnetic moments by analyzing the difference between an X-ray absorption spectra taken with a positive (σ^+) circularly polarized light, and that taken with a negative (σ^-) one, when applying a magnetic field (here, $B = 5$ T). This difference spectrum represents an X-ray magnetic circular

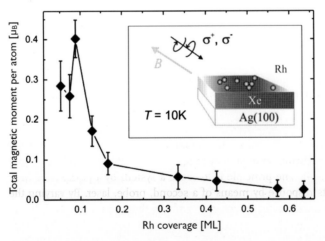

Figure 5.5 Total moments per atom in Rh clusters deposited on a Xe buffer layer at $T = 10$ K. The inset is a sketch of the experimental setup where X-ray magnetic circular dichroism (XMCD) has been used to measure magnetic moments (see text for details). Adapted from [125].

dichroism (XMCD). The analysis of such an XMCD spectrum gives access to the magnetic properties of the system, such as its spin and orbital magnetic moment. The basic trends of the curve plotted in Figure 5.5, with a maximum and a decreasing asymptotic behavior, are well understood with standard theories of magnetism, on the basis of modifications of the electronic density of states by local coordination effects. The moments decrease with Rh coverage (in a monolayer unit), due to the increase of the average cluster size. The increased magnetism for moderate sizes, in turn, is related to crystal field effects to the extent that the moments are no longer quenched if the cluster size has sufficiently decreased. Another interesting aspect of these results is the importance of decoupling between Rh clusters and substrate for the observation of magnetism, an effect which is not yet fully understood from the theoretical point of view. This points to the fact that there still remains a lot to be done in future experiments with greater structural control, such as atomic manipulation, in order to study emergent magnetism in Rh or other nanostructures.

5.1.3
From Observation to Manipulation

5.1.3.1 Helium Droplets as a Laboratory

As pointed out in the previous section, contact of cluster with the environment can induce a mutual modification of both materials. On the other hand, substrates are very useful to handle and control clusters (temperature, orientation) to an extent which is not easily possible for free clusters. This calls for setups in which the impact of the environment is as controlled/reduced as possible. Helium is the least interacting material and thus droplets of liquid He constitute the ideally inert confining device [126]. Practically, one places the cluster in a bubble within a He droplet, for instance by using pick-up sources, see Section 2.1.1. The He droplet allows to isolate the embedded cluster and to control parameters such as the temperature. Indeed we know that it may strongly affect cluster properties (see for example the discussion of Section 4.2.3). One may then perform a large variety of experiments on such an "isolated" cluster, for example, scanning various dynamical regimes. As an example, we mention again that the optical responses of embedded Ag clusters in Figure 5.1 were obtained in such a setup. In that case, the Ag_8 and its rare gas coating was itself embedded into a He droplet.

More violent scenarios can also be considered as in the example of Figure 5.6 which employs embedded Ag clusters as a tool for a pump and probe study. Such experiments are a particular case of analyzing clusters with dedicated pulse profiles (see Section 4.4.4). The principle is to excite a system by a pump laser first, and then explore its response by means of a second, probe, laser. By varying the delay between probe and pump pulses, one explores the various stages of the response of the system and thus collects a "time-resolved" picture of the dynamical response [26]. Figure 5.6 shows the yield of Ag^{5+} ions as a function of pulse delay for two different intensities. Each case clearly shows an optimal delay time where the yield is maximal. This reflects the subtle interplay between cluster expansion and resonance conditions. The laser frequency of 1.54 eV (identical for the pump

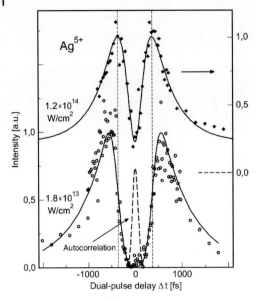

Figure 5.6 Yield of Ag^{5+} ions from the Coulomb explosion of large Ag clusters embedded in He clusters, from two subsequent laser pulses as a function of the delay time between the pulses. Each pulse has a frequency of 1.54 eV and a pulse length of 130 fs. Two intensities are considered as indicated. From [128].

and the probe) is safely below the plasmon resonance energy in the initial Ag cluster (typically about 4 eV). The moderate ionization by the first (pump) pulse triggers a slow cluster expansion and a corresponding decrease of the plasmon frequency (see Section 2.2.3). At a certain time, the plasmon comes into resonance with the laser. This, in turn, leads to a large energy absorption with subsequently violent atomic ionization processes. The amount of ionization from the first pulse of course depends on the laser intensity: less intensity produces less first ionization, thus slower expansion, and with it, a longer time span until resonance conditions are reached. That is precisely what is seen in Figure 5.6: the case with smaller intensity (lower curve) has its maximum at a larger delay time than the case with higher intensity (upper curve). This interpretation has been corroborated by dynamical simulations [127].

5.1.3.2 Tailoring Cluster Color

The interplay with a substrate provides various possibilities to tailor cluster properties and this opens a wide field for applications. An example for dedicated construction of cluster color is given in Figure 5.7. It shows the spectra of an array of small Ag_2S clusters produced and kept in the cavities of Ca zeolite [129]. The latter one is a microcrystal based on aluminum silicate, which provides a regular porous structure. Clusters are grown and kept in the small cavities of the zeolite crystal. This establishes, in fact, an array of embedded $(Ag_2S)_N$ clusters. The largest cluster which fits into a cage is $(Ag_2S)_4$. The average size of the clusters is determined by

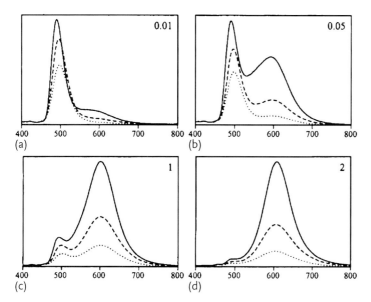

Figure 5.7 Photoluminescence spectra drawn versus wavelength (in nm) for Ag_2S clusters in the cavities of Ca zeolite. The average number of Ag^+ ions per cavity are given as 0.01 (a), 0.05 (b), 1 (c) and 2 (d). These numbers then determine the size of the Ag_2S clusters and their density over the zeolite cages. Spectra are taken at three temperatures: 78 K (solid lines), 173 K (dashes), and 223 K (dots). After [129].

the supply of Ag^+ ions during growth. Figure 5.7a shows the smallest Ag content where Ag_2S molecules prevail. The luminescence is dominated by a blue-green line corresponding to an electronic dipole transition in the molecule. With increasing cluster size (see Figures 5.7b–d), this molecular line is slowly superseded by a broad line in the range of orange-red light which finally takes over. The orange-red luminescence is most probably caused by Ag_8S_4 clusters. A fully microscopic description of these trends has not yet been achieved because the $(Ag_2S)_N$ clusters possess an enormous number of isomers, not to mention the complications added by including the interface interaction. One furthermore needs to account for the mode coupling in the array of clusters, as discussed in the paragraph below, which is probably responsible for the broadening of the observed cluster emission line. The interesting aspect for practical applications is that cluster properties, as here the average cluster size and with it the color, can be easily tuned by the environment and the chemical conditions during cluster growth.

5.1.3.3 Coupled Plasmons

We have seen at many places in this book that metallic nanoparticles have a pronounced plasmon resonance (see Sections 2.2.3 and 4.2). Ensembles of them thus strongly interact through the dipole–dipole interaction mediated by plasmon response. They hence constitute a system of coupled oscillators showing all effects of mode coupling. The rich world of thus emerging scenarios is reviewed in [130].

Such arrays of coupled nanoparticles are much discussed in technical applications of clusters.

Figure 5.8 sketches two examples for the simple case of two coupled modes. Figure 5.8a addresses the case of a "molecule" of two metallic nanoparticles staying at a certain distance from each other. The (identical) plasmon energies of the single clusters are indicated on the left and on the right spectrum. The angular momentum $l = 1$ stands for the Mie surface plasmon. There also exist surface plasmons at higher l from which the next important case $l = 2$ is also indicated. The two nanoparticles are coupled by a dipole–dipole interaction which reads as:

$$V_{\text{dip}} = e^2 \frac{D_1 \cdot D_2 - 3 D_1 \cdot e_{12}\, e_{12} \cdot D_2}{R_{12}^3}, \quad e_{12} = \frac{R_{12}}{R_{12}}, \tag{5.1}$$

where R_{12} is the distance vector between nanoparticle 1 and nanoparticle 2. This coupling removes the degeneracy of the isolated modes and produces two modes, one lower and another one higher. The energetic splitting is of order of interaction strength which decreases as R_{12}^{-3}. Figure 5.8a indicates by dotted lines that there may also be some mixing with modes of higher l because the rotational symmetry is broken such that angular momentum ceases to be a good quantum number. The extension to larger groups of interacting nanoparticles is obvious.

Figure 5.8b shows a somewhat surprising scenario of two coupled modes which reside in one and the same system. Consider a metallic nanoparticle with a large cavity inside, for simplicity both spherical, particle and cavity. The cavity may be vacuum or some insulating material. The resonance spectrum of this system can be constructed by separating it into two simpler subsystems, a compact nanoparticle as indicated (left side) and a spherical cavity in bulk (right side). Both have a pronounced Mie plasmon resonance. Dielectric theory yields the frequencies as indicated. These two modes are now coupled by a dipole–dipole polarization force which then yields the two resonances shown in the middle. As expected for cou-

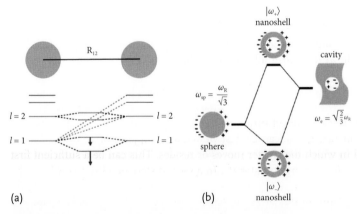

Figure 5.8 Schematic view of coupled plasmon modes. A "molecule" of two metallic nanoparticles (a) and a spherical metallic nanoparticle with a spherical cavity inside (b). Adapted from [130].

pled two-level systems, we find one mode shifted below the smaller frequency of the isolated system, and another mode shifted above the higher frequency. The actual strength of the shift depends on the details of the setup, for example, the thickness of the metallic layer and the dielectric constant of the cavity material.

5.2 On the Description of Embedded/Deposited Clusters

5.2.1 Brief Review of Methods

5.2.1.1 Towards Hierarchical Approaches

As mentioned in previous sections, combined cluster plus environment systems are much more complicated in their description than merely free clusters. The interface particularly requires a very careful treatment. The conceptually simplest procedure is to use the same (detailed) approach for cluster and environment. In such calculations, the substrate is represented by rather small pieces of material kept structurally close to the known surface and bulk configuration. An alternative strategy is to freely vary the structure of the composite systems as a model for clusters in/on environment. Both strategies have their weaknesses due to the extremely limited number of substrate atoms. Figure 5.9 shows a typical configuration as used for such full quantum mechanical calculations of cluster and substrate, here for a Na cluster on a Cu surface. Since each Cu atom carries 11 active electrons, these calculations are obviously rather expensive, in spite of the fact that the representative of bulk material is rather small.

At the other extreme, very simple theoretical approaches have been developed relying on a macroscopic description of the substrate as jellium or by dielectric theory, see for example [12]. Such a macroscopic approach is valuable for first explorations of trends and orders of magnitude. But it becomes clearly insufficient when detailed structures play a role or large-scale ionic motion sets in. The latter one may still be described by simple molecular dynamics (see, for example, Figure 3.1). This however misses all quantum effects which are important for cluster electrons (see Section 4.1.1).

If one wants to describe the excitation and dynamical response of realistic systems with large numbers of degrees of freedom, one needs to find a compromise between approximate and yet sufficiently detailed description. An inert substrate, as insulators or rare gases, can be first viewed as a supplier of an external interface potential in which the cluster moves or resides. This can be a sufficient first order approach for very inert materials. In a second step, one has to envision the static and dynamic response of the substrate itself (see Figure 5.2). The third step would be to account for the hybridization of electronic states between cluster and substrate. This step can be omitted if we deal with electronically inert, insulating substrates. Steps 1 and 2 considered together suggest an approach which takes care of the interface potentials and the polarizability of the environment, while al-

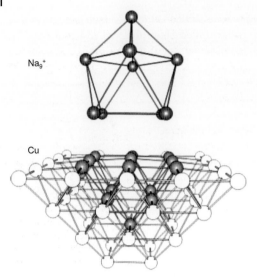

Figure 5.9 Typical configuration for a DFT calculation of Na$_9^+$ deposited on Cu surface, treated at two levels of refinement. Atoms indicated by filled circles employ a large basis set for the electronic calculation and the ionic positions are freely varied. Atoms indicated by open circles are frozen at crystal positions and their electrons are treated at lower resolution. Adapted from [131].

lowing a lower level description in other aspects of the material. The idea is thus to start from a fully microscopic description of the cluster and to treat the environment at a simpler level, namely as classical particles with an internal dipole polarization, also handled as classical degrees of freedom. This point of view naturally defines a hierarchy amongst the various degrees of freedom. Such hierarchical approaches are widely used in many complex systems and is often called quantum mechanical/molecular mechanical modeling (QM/MM), for a general overview of the method, see [132].

5.2.1.2 The Basics of QM/MM Modeling

As suggested above, a QM/MM approach is based on the partition of a system into two pieces, an "inner" piece being treated at a high level of microscopic detail (usually an elaborate quantum mechanical approach, QM) and an "outer" shell treated at a lower level of detail (usually on the basis of molecular mechanics, MM). The total energy is thus written as

$$E_{total} = E_{QM/QM} + E_{QM/MM} + E_{MM/MM} , \qquad (5.2)$$

where the three terms respectively represent the fully QM contribution (first term), the fully MM one (third term) and the interaction energy between QM and MM pieces (second term). The typical inputs for the QM/QM term are the electronic kinetic energy, the ions (nuclei)–electrons attraction, the electron–electron repulsion and the ion–ion repulsion. The MM/MM term includes all typical molecular dynamics terms (see Section 3.1.5) including elongation, deformation torsion of molecular bonds, as well as van der Waals interactions between MM atoms and electrostatic charge-charge effects if needed. Finally, the QM/MM interaction part accounts for electrostatic interactions between electrons or ions and charges, and van der Waals interactions between QM and MM atoms. The account of the MM part, and especially of the QM/MM interaction, is essential as it allows polarization effects from the MM environment onto the QM wave function, an effect which is usually long range.

There exists a wide variety of QM/MM approaches depending on the employed QM and MM descriptions. The QM may be described by any of the standard approximations for treating the electronic problem in clusters or molecules (see Section 3.2). The MM models differ by the classical force fields one uses, which sensitively depend on the considered systems (see Section 3.1.5). Moreover, the MM system may be further divided into layers which are treated at different levels of refinement down to a simple dielectric response. Finally, the interface between QM and MM can be modeled in different ways. The most important distinction still lies at the first step, the separation into QM and MM subsystems (and possible subdivision therein). The optimal choice is not always obvious in general as discussed above. But a division line is naturally given in the case of clusters on/in inert substrates. We will illustrate a detailed QM/MM modeling for one typical example in Section 5.2.2.

5.2.1.3 The ONIOM Approach

One particular variant of QM/MM modeling consists of the ONIOM approach. ONIOM stands for "our own N-layered integrated molecular orbital and molecular mechanics." This approach also requires a division into QM and MM parts (again with possible subdivisions therein), but it avoids the often cumbersome modeling of the QM-MM interface [133]. As an illustration, we consider the case of ONIOM–2 where the system is plotted as two pieces, that is QM and MM without further subdivisions. One separates the treatment into total system $\mathcal{R}$ ($\equiv$ QM + MM) and the "model" system $\mathcal{M}$ ($\equiv$ QM). As a further distinction, one distinguishes the treatments into high level (H, typically a QM calculation) and low-level (L, typically a MM calculation) ones. The total energy is then composed as

$$E_{\text{total}} \simeq E_{\mathcal{R}}^{\text{ONIOM-2}} = E_{\mathcal{M}}^{\text{H}} + E_{\mathcal{R}}^{\text{L}} - E_{\mathcal{M}}^{\text{L}} . \tag{5.3}$$

The major difference and simplification as compared to the above discussed QM/MM approaches lies in the absence of explicit treatment of the interactions between the QM and MM parts of the system. This however requires that one disposes of a readily developed and reasonably well-working low-level approach for the whole system $\mathcal{R}$. As with any QM/MM modeling, the ONIOM approach is obviously adapted to cases where the separation between the two parts of the system is "natural" as for example a molecule ($\mathcal{M}$) inside a solvent ($\mathcal{R}$ being then the molecule plus the solvent). Still ONIOM also works in many other situations even when the system has to be split "artificially" into QM and MM parts. Figure 5.10 illustrates the capability of the method [134]. It compares a DFT approach for the full system ($\mathcal{R}$) and an ONIOM model to experimental crystallographic structure

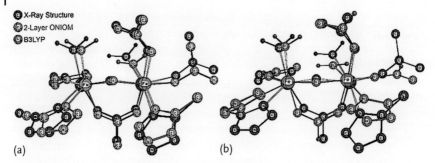

Figure 5.10 Superimposing of experimental X-ray structure of an enzyme (darker gray) and optimized geometries calculated with the 2-layer ONIOM (a) and DFT (b). From [134].

of a large molecule. At least in that case, the approximate ONIOM treatment even improves the agreement with experiment as compared to the pure DFT modeling, since the ONIOM structure better reproduces the experimental one, as is visible in Figure 5.10a.

5.2.2
An Example of QM/MM for Modeling of Deposited/Embedded Clusters

As previously mentioned, the case of embedded/deposited clusters is calling for hierarchical methods in the spirit of QM/MM. Moreover, it is a situation for which a QM/MM approach is especially well suited, as the separation between the QM and MM parts is rather obvious. Still, even if the separation problem is to a large extent naturally solved, a proper modeling requires to carefully account for the interactions between QM and MM subsystems. This is especially true when metals are involved since they are very sensitive to polarization effects. This aspect becomes even crucial if one considers charge effects which occur, for example, when irradiation processes lead to ionization. As a typical example, one can consider laser irradiation of a metal cluster deposited on an insulating surface. This is the situation we want to exemplify in this section and analyze in a bit more detailed manner. Note that the insulating nature of the surface (even if polarizable) is favorable here as it justifies an even better separation between QM and MM parts. For simplicity of the presentation, we have chosen the inert Ar surface.

5.2.2.1 The Degrees of Freedom
We describe the cluster with the same methods as free clusters, see Chapter 3. Here, in particular, we use TDLDA for the electrons coupled to molecular dynamics (MD) for cluster ions. The Ar atoms of the substrate are treated classically, attributing two ($\times 3$) degrees of freedom to each: position and dipole momentum.

Table 5.1 Degrees of freedom (DOF) of the hierarchical model of a Na cluster in contact with an Ar substrate. The first two rows are for Na DOF, and the last two ones are for Ar DOF.

DOF	Counter	Description
$\varphi_i(\mathbf{r})$	$i = 1, \ldots, N_{el}$	Single-particle wave functions for Na valence electrons
$\mathbf{R}_{I^{(Na)}}$	$I^{(Na)} = 1, \ldots, N_{Ion}$	Positions of the Na$^+$ ions
$\mathbf{R}_{J^{(c)}}$	$J^{(c)} = 1, \ldots, N_{(c)}$	Position of the Ar cores
$\mathbf{R}_{J^{(v)}}$	$J^{(v)} = 1, \ldots, N_{(v)}$	Position of the Ar valence clouds

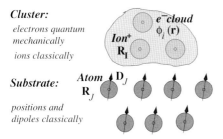

Figure 5.11 Sketch of the hierarchical quantum mechanical/molecular mechanical (QM/MM) model for metal clusters in contact with a rare gas substrate. Actually, atom position $\mathbf{R}_J$ and dipole momentum $\mathbf{D}_J$ are handled in terms of core and valence cloud.

The latter allows to account for polarization effects at the side of the substrate. Due to active dipoles, polarizability is treated fully dynamically and can thus account for the substrates dynamical response to external fields, particularly those created by the cluster. The dipole momentum is described, in practice, by associating to each atom a core and a (rigid) valence shell with opposite charges. A difference between core and valence position naturally generates the dipole momentum. We label in the following these two degrees of freedom by $J^{(c)}$ for cores and $J^{(v)}$ for valence clouds of the MM (here, Ar) atom J. Table 5.1 summarizes the various degrees of freedom and Figure 5.11 complements that by a graphical illustration of the setup. Only a few Ar atoms have been sketched to keep the figure simple but one actually deals with hundreds of atoms. These are arranged in (4–8) layers with periodic boundary conditions in the lateral direction to effectively simulate infinite layers, while the lowest layer is frozen at the positions of bulk crystal [135]. This automatically places the layers above in nearly crystal structure (here, an fcc lattice) and yet allows for surface deformations through the deposited cluster. Without the deposited cluster, we obtain the (001) surface of Ar crystal.

5.2.2.2 Modeling of the Hamiltonian

The starting point is the total energy in typical QM/MM form, see Eq. (5.2):

$$E = E_{Na} + E_{substr} + E_{coupl}, \tag{5.4}$$

where E_{Na} describes an isolated Na cluster, E_{substr} the substrate, and E_{coupl} the coupling between the two subsystems. The Na cluster is described at the DFT level (see Section 3.2.4).

The energy of the substrate subsystem is given by

$$E_{substr} = \sum_{\tau, J^{(\tau)}} \frac{M_\tau}{2} |\dot{\mathbf{R}}_{J^{(\tau)}}|^2$$
$$+ \frac{1}{2} \sum_{(\tau, J^{(\tau)}) \neq (\tau', J^{(\tau')})} V_{\tau\tau'}(|\mathbf{R}_{J^{(\tau)}} - \mathbf{R}_{J^{(\tau')}}|)$$
$$+ \sum_{J^{(c)} = J^{(v)}} \left[\frac{K}{2} (|\mathbf{R}_{J^{(c)}} - \mathbf{R}_{J^{(v)}}|)^2 - V_{cv}(|\mathbf{R}_{J^{(c)}} - \mathbf{R}_{J^{(v)}}|) \right], \quad (5.5)$$

where $\tau \in \{c, v\}$. The M_τ and q_τ are masses and charges associated with component τ. These model parameters are to be calibrated to basic properties and interactions of the corresponding subsystem. The second sum in the first line refers to interaction potentials between different atoms. They consist of two terms, namely:

$$V_{\tau\tau'}(r) = q_\tau q_{\tau'} e^2 \frac{\text{erf}(r/\sigma)}{r} + f_{\tau\tau'}^{(\text{short})}(r). \quad (5.6)$$

The first term with the error function, $\text{erf}(x) = 2/\sqrt{\pi} \int_0^x dy \exp(-y^2)$, represents the long-range Coulomb (polarization) interaction and is the same for any constituent. The second term stands for the short-range interaction and is mostly repulsive. In particular, it accounts for Pauli repulsion from atomic cores and interaction effects besides Coulomb. Its form and parameterization sensitively depend on the actual system [135]. The third line in Eq. (5.5) models the interaction between core and valence cloud in the same atom by an harmonic oscillator. Note that the core-valence potential V_{cv} is subtracted to avoid double counting.

The coupling energy is similarly given by Coulomb interaction and some short-range repulsion, as:

$$E_{coupl} = \sum_{\tau, J^{(\tau)}} \left[\int d\mathbf{r} \rho_{el}(\mathbf{r}) V_{\tau, el}(|\mathbf{r} - \mathbf{R}_{J^{(\tau)}}|) + \sum_{I^{(Na)}} V_{\tau, Na}(|\mathbf{R}_{J^{(\tau)}} - \mathbf{R}_{I^{(Na)}}|) \right] + E_{VdW}, \quad (5.7)$$

where the interaction potentials $V_{\tau,el}$ and $V_{\tau,Na}$ are also given in the general form (5.6). The last term is a place-holder for a possible van der Waals (VdW) energy between cluster electrons and substrate atoms. It is an important contribution for interaction with rare gas atoms but usually negligible in ionic crystals such as MgO or NaCl.

The various interaction potentials introduce several model parameters which are taken from literature or fitted from binary system data. Possibly pending parameters are finally fitted to *ab initio* calculations of larger compounds. In the case we presently discuss, the Ar–Ar interaction is taken from the literature and other parameters are calibrated from data for the Ar–Na$^+$ and Ar–Na dimers.

The total energy (5.4) determines the dynamical equations by variation. This proceeds analogous to the case of free clusters, see Chapter 3, where the pool of classical particles has been extended to the substrate degrees of freedom. Therefore we can also take over from the case of free clusters, treated at the level of TDLDA-MD, the numerical methods to solve these coupled equations.

5.2.3
A Few Typical Results

5.2.3.1 Mechanisms of Optical Shift

As a first application, we analyze the impact of the environment on the optical response in a way similar as already discussed in Figure 5.1 but this time using our model Na@Ar system and trying to better understand underlying physical mechanisms. With the strong polarization interaction from the substrate, one indeed expects large effects on the cluster's optical response. Figure 5.12 shows the optical absorption strength for Na_8 under varied conditions, free cluster (dashed), embedded in an Ar_{164} matrix (solid line), and again embedded but using only the (repulsive) short-range potentials in the cluster-substrate coupling (dotted line). Na_8 has a closed electron shell and thus is highly symmetric. It thus displays a single plasmon peak (see Section 4.2.2), even when embedded. The peak resides around 2.6 eV for the free cluster. For the embedded one, we consider two stages. First, we only take into account the Ar core repulsion, see dotted curve. This strongly blue-shifts by about 0.4 eV the plasmon peak, which is now located at 3 eV. This reflects the "compression" exerted by the matrix on the Na cluster. It is mostly a short-range effect at the interface between cluster and matrix, little affected by the size of the environment. Next, we step up to the full description by activating the Ar dipole degrees of freedom and the VdW interaction in the calculation (solid line). That leads to a large red-shift which brings the final peak back down to 2.75 eV. This polarization effect is of long range (at variance with the above mentioned blue-shift). Eventually, we see only a moderate shift from free to embedded due to a cancella-

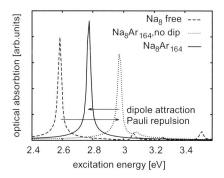

Figure 5.12 The Mie plasmon peak frequency for free Na_8 (dashed line) and for Na_8 embedded in an Ar_{164} matrix (solid line). For the latter case, we also show a calculation without dynamical dipole response of the Ar atoms in the matrix (dotted line).

tion of two large effects, that is a blue-shift by short-range repulsion and a red-shift by long-range polarization potentials. Similar compensating effects are seen everywhere in such combined cluster-substrate systems. The results of such compensations is hardly foreseeable by simple models. One needs careful modeling and calibration to achieve nearly reliable predictions. These compensation effects act similarly in the case of the Ag cluster coated by rare gas atoms (see Figure 5.1), also leading here to the same subtle net effects.

5.2.3.2 Structure and Laser Irradiation

Figure 5.13 shows results for other observables in the case of deposited Na_8 but on another substrate, namely MgO. The latter, being an insulator, is also electronically inert and can be well treated as the MM part in a QM/MM approach. The modeling is more involved than the Ar example as two different species are involved.

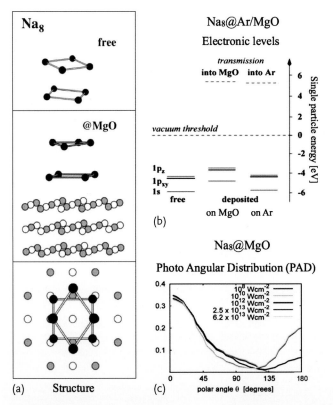

Figure 5.13 Illustration of the effects of deposition for the case of Na_8@MgO. The structure of the system (showing only a small part of MgO substrate) is presented (a). Also shown is the change of single-electron spectra, particularly of the IP, through the substrate (b) and the effect on the angular distribution of electron emission following excitation by a laser with pulse length 25 fs, polarization perpendicular to the surface, varying intensity for a fixed frequency $\omega_{las} = 4.76$ eV (c). Angles are measured with respect to the laser polarization, that is with respect to the normal to the surface.

Figure 5.13a illustrates the structure of the system. The Na$_8$ cluster is gently tied to the substrate by polarization forces. The structure, consisting of two rings of four ions twisted by 45° to minimize Coulomb interaction, remains very close to that of the free cluster (upper left). There are some perturbations of the position of the Na$^+$ ions (when Na$_8$ is deposited) and of the nearest substrate ions which, however, are too small to be visible in a plot.

Figure 5.13b shows an observable from the cluster electronic structure, namely the single-particle energies of Na$_8$ under different conditions, free, deposited on MgO, and deposited on Ar surface for comparison. The electronic structure of the occupied levels is arranged the same way in all three cases: a deeper bound 1s state and a nearly degenerated group of 1p states above that. The energy difference between the two groups of levels also remains about the same. What changes the most is the overall binding, the energy of the bound state relative to the continuum threshold, and consequently the IP of the system. We see a considerable mellowing of binding for MgO but a nearly negligible one for Ar. The reason is that the binding of the cluster to Ar surface is generally softer, resulting in a larger distance between cluster and substrate which, in turn, only leaves a very small effect on the electronic states. Also shown are the thresholds for electron transmission into the substrates. These are rather high and about the same. This threshold lies at the lower end of the conduction band and the high value demonstrates that MgO as well as Ar are good insulators.

Figure 5.13c shows photoangular distributions (PAD) of electrons emitted after irradiation with a strong laser pulse of fixed frequency and varying intensity. The laser polarization is perpendicular to the surface. Distributions are drawn versus angle θ relative to the laser polarization axis, hence normal to the surface. Therefore $\theta = 0$ and $\theta = 180°$ stand for emission perpendicular to the surface. All distributions are strongly normal to the surface. For low intensity, they are mainly forward peaked ($\theta = 0$) and emission into the substrate ($\theta = 180°$) is much suppressed. Higher intensities allow photoelectrons to penetrate into the substrate and to develop a significant PAD around $\theta = 180°$. This has to be compared with a free cluster where electron emission is symmetric in $\theta = 0 \leftrightarrow 180°$. With this dynamical observable of electron emission, we encounter a case where the effect of the substrate is dramatic.

These examples show that every outcome is possible for the electronic properties, from minor effects due to weak attachment or compensations to dramatic changes through the substrate. One needs to carefully consider each combination of materials and each observable separately.

5.3
Clusters and Nanosystems

Clusters are nanoparticles and as such, they belong to the highly celebrated realm of nanophysics with its expected high potential for applications in technology. An entire book could be devoted to all aspects of clusters in nanoscience. This goes

5.3.1
Towards More Miniaturization

One important line of development is the miniaturization of electronic switching devices. In this direction, carbon nanotubes have been found to be promising candidates for a minitransistor. One can go even smaller down to one single C_{60} [136], as demonstrated in Figure 5.14. The current-voltage characteristics sensitively depends on the gate voltage V_g, in particular the conduction gap (bias voltage range of zero I). At the end of the gap, conduction then sets on stepwise. This is due to quantization of charge (one single electron is involved) and the voltage where the step occurs. This is related to the energy cost of adding or removing one electron to the C_{60}.

Taking a closer look, we spot a small, second step about 5 mV above the first step. It can be worked out that this small step is related to one quantum of the center-of-mass oscillations of the whole C_{60} in the weak binding potential (of van der Waals type) between the electrodes [136]. Thus this setup is more than a transistor. It can also constitute a nanomechanical device which thus connects with another field of great actual interest, namely nanomechanics [137]. Although most nanomechanical devices are carved from bulk material, clusters may become increasingly important in this field, as the above example shows. It also allows to demonstrate the various levels of theoretical description. The gross features of the jumps in charge transfer could be estimated from properties of free C_{60}. A refinement is possible

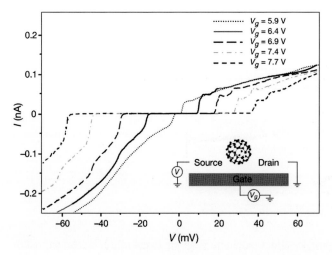

Figure 5.14 Current I versus bias voltage V for a single C_{60} placed between two electrodes (see insert). Curves are shown for different gate voltages V_g. Adapted from [136].

by accounting for the mean fields from the electrodes where, in particular, the van der Waals force was found to be decisive. This complies with hybrid treatment presented in Sections 5.2.1 and 5.2.2. A fully microscopic description of conductance is the most demanding step as this requires a microscopic modeling, not only of the C_{60} but also of the electrodes, see Section 3.4.3.

5.3.2
On Catalysis

Catalysis was one of the strong motivations for the development of cluster physics in particular because metallic nanoparticles are often found to be efficient catalysts [138]. This especially holds for Au clusters, see for example [139]. The case is, however, not that simple. The catalytic activity sensitively depends on details as size, shape or environment. We will discuss here three typical examples.

Figure 5.15 demonstrates the effect of a substrate on the catalytic activity of Au_{20} clusters. The starting configuration is Au_{20} clusters on an MoO surface (uppermost curve). The curve is flat which means that no catalytic effect is seen in that case. Then one adds stepwise an MgO film on the MoO substrate. The average number of monolayers (ML) of MgO is indicated on each curve. Already the first ML points out a catalytic activity with a wide bump in the catalytic yield of $^{13}CO_2$ around 300 K. This activity is shifted gradually to lower temperatures (down to about 250 K) with increasing number of MgO layers. Changes are seen surely beyond the third ML. This indicates that very subtle effects on the electronic configuration have a

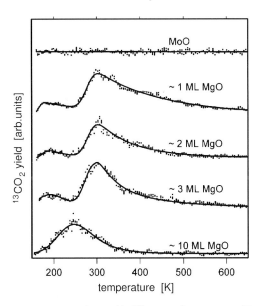

Figure 5.15 Catalytic yield of $^{13}CO_2$ in the presence of Au_{20} clusters on different substrates, on pure MoO and on various monolayers (ML) of MgO on MoO substrate as indicated. The curves are shifted relative to each other for better discrimination. Adapted from [140].

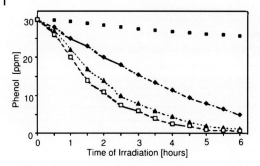

Figure 5.16 Time evolutions of decomposition rate of phenol in sunlight by using different concentrations of Au clusters in TiO$_2$ film. Compared are pure phenol (filled boxes), phenol on TiO$_2$ (diamonds), on 1% Au/TiO$_2$ (triangles), and on 2% Au/TiO$_2$ (open boxes). Adapted from [142].

large impact on catalysis. This also indicates that a pertinent description of catalytic properties by theoretical simulations is certainly extremely demanding. On the other hand, this great sensitivity is very promising for the future improvement of catalytic efficiency.

Not only the substrate, but also the concentration of clusters on the substrate can have a large effect on catalytic activity. This is demonstrated in Figure 5.16 for the case of photocatalysis of phenol (for a review, see [141]). The time evolution of the decomposition rate of phenol in sunlight, using different concentrations of Au clusters in a TiO$_2$ film, is plotted. The temporal profile is qualitatively similar in all cases. But the slope, that is the catalytic activity, dramatically differs. It is obvious that the larger content of Au nanoparticles enhances the decomposition rates.

It is moreover known that the catalytic reactivity of small noble metal clusters sensitively depends on size [114]. This is not surprising as the electronic structure which is responsible for the reactivity can dramatically change from one cluster size to the next (see, for example, Section 4.1.1.1). The size effect fades away together with electronic shell effects for very larger clusters of diameter in the range of several nanometers. However, these large systems can appear in substantially different geometries. These can produce, of course, very different electronic surface states which, in turn, has a large impact on catalysis. Figure 5.17 shows an example for the case of Ag nanoparticles [143] (see also the review [144]). By varying the initial concentration of a Ag salt and other reaction conditions, it is possible to produce Ag nanoparticles in three very different geometries: triangular plates (Figure 5.17a), nearly spherical polygons (Figure 5.17b), and cubes (Figure 5.17c). These geometries have been identified by scanning electron microscopy. The schematic plots in Figure 5.17d–f illustrate the geometries and dimensions. These three configurations have been exposed to styrene and the rate of catalytic conversion was accumulated over several hours. The resulting net rate is shown in Figure 5.17g. The differences in rates are dramatic. The cubic configuration is more than a factor 14 more efficient than the triangular plates. The effect can be related to electronic properties of surface states for the different boundary conditions. This requires further expla-

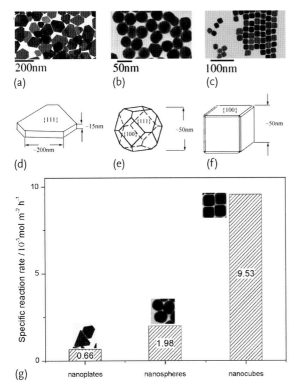

Figure 5.17 Shown are SEM images of Ag nanoparticles in different shapes: triangular plates (a), near spherical (b), and cubic (c). Also represented schematically are the corresponding structures (a related to d, etc.) of the three differing shapes (d–f). The plotted data presents the catalytic reaction rates for conversion of styrene measured over 3 hours (g). Adapted from [143].

nation of surfaces. One can cut a fcc crystal in different ways which, subsequently, displays different surface structure. In Figure 5.17, we are distinguishing between the (100) surface, which is obtained by cutting the fcc metal parallel to the layers of the fcc unit cell and the (111) surface which is obtained by cutting the fcc metal along a plane orthogonal to the diagonal between the x, y and z axes. It is expected that the (100) surfaces of the nanoparticles show more catalytic reactivity than the (111) surfaces. The cube has indeed the largest fraction of (100) surfaces and thus is the most reactive [143].

With all these subtle influences on catalysis (sizes, combination with substrate, geometries), one sees that the topic of clusters and catalysis still provides a rich field for research and applications. There are most likely a couple of promising configurations which are still awaiting discovery.

5.3.3
Metal Clusters as Optical Tools

Metal clusters are distinguished by a particularly strong optical response near the Mie plasmon frequency (see Sections 2.2.3 and 3.4.4). This is associated with a strong, resonant near-field enhancement [145]. The effect can be exploited in various ways. We take here an example from [146] dealing with accelerating the fluorescence of a terrylene molecule by Au nanoparticles. Terrylene belongs to the polycyclic aromatic hydrocarbon (PAH) family, a kind of molecule very much studied in an astrophysical context (see Section 6.2). The experiment is performed in a multilayer setup as sketched in Figure 5.18a. The effect of neighboring Au clusters is obvious. They lead to more intense photon emission (Figure 5.18b) and thus shorter fluorescence lifetimes (remember the magnification by 1000 and 10 applied on the top and middle curves). The presence of a single Au_N close to the terrylene already delivers enhancement factors of 3–28. Also, placing the molecule in between two clusters makes a factor as large as 111. The enhancement may come from the enlarged extension of the dipole antenna yielding a larger dipole matrix element. But the plasmon coupling also changes the density of states. This can furthermore enhance the emission if a higher density lies in the appropriate frequency range (for effects of coupled plasmons, see also Section 5.1.3). These vague arguments indicate that a quantitative prediction requires detailed calculations. It is to be noted that the sample molecule, here terrylene, as such is already a particularly efficient emitter. Even larger gains are expected for molecules with lower quantum efficiency.

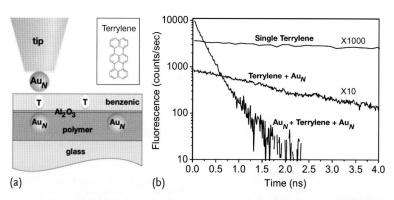

Figure 5.18 (a) Schematic view of the experimental setup. From bottom to top: glass substrate, Au nanoparticles embedded in a polymer matrix, thin film of Al_2O_3, benzenic film containing low concentration of terrylenes (T), eventually an Au nanoparticle on top of a terrylene, and tip for the fluorescence measurements (at 532 nm). The insert shows the structure of the terrylene molecule. (b) Fluorescence signal as a function of time for a single terrylene molecule (top curve, ×1000), a terrylene in close vicinity of a single Au cluster (middle curve, ×10), and a terrylene sandwiched between two Au clusters (bottom curve). Adapted from [146].

5.3.4
Composite Clusters and Nanomaterials

There is a proposal to make use of the structural properties of composite clusters to design new nanomaterials by using particularly well-bound clusters as "superatoms" which can bind to large molecules or even solids made of superatoms. The elementary building block, being in itself a complex system, provides a rich scenario of couplings to other building blocks of the same type. This can be exploited to design materials with wanted properties. There is, of course, a world of possibilities for such building blocks, or superatoms. Fullerenes are particularly well bound (by cooperation of electronic plus atomic shell closures) and an ideal building block for designing materials as chains, arrays, or other geometrical compositions [147].

To indicate the richness of the options, we show in Figure 5.19 a less prominent example of composites from superatoms, namely designer molecules assembled from $(Al_{13}K_3O)$ clusters [148]. The key point here is the fact that the particular cluster $(Al_{13}K_3O)$ is much more strongly bound than other Al clusters (with and without attachment of other elements), namely a binding of 5.53 eV for $(Al_{13}K_3O)$, as compared to a binding of 2.51 eV for the closest comparable object $Al_{13}K$. Thus it is conceivable to combine several $(Al_{13}K_3O)$, while maintaining the superatom's identity. Figure 5.19 shows a few theoretically estimated examples for stable configurations of such composites [148]. The superatom consists of a Al_{13} cluster with one K_3O attached at one side. This defines a systems axis along which the composites line up. They weakly line up and thus, eventually ring structures develop. The example demonstrates that such constructions live from a difference between strong binding of a particular cluster (= superatom) and the weaker binding forces

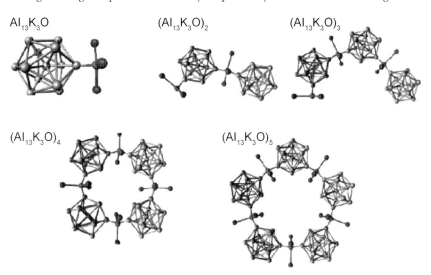

Figure 5.19 Configurations of composites from material design by various $Al_{13}K_3O$ building blocks. Adapted from [148, 149].

between the clusters which, additionally, can have pronounced nonisotropic structure and allow some bending in larger units. It is the hope that this way of versatile constructions become possible when exploiting these cluster properties in a creative manner.

Another huge field for applications of clusters or assemblies thereof is the design of nanosensor devices. As most examples of these applications concern the tracking of organic molecules and biological environment, we will specifically discuss this topic in Section 6.4.

6
Links to Other Areas of Science

As we have seen in the previous chapters, cluster science has emerged as a major field of research over the past few decades. Clusters are fascinating objects interpolating between atoms/molecules and bulk and exhibit marked finite size effects, as exemplified at various places. We have also seen that cluster properties may be strongly altered by the presence of an environment and that environment properties themselves may be affected. But cluster science offers many further aspects as it also has a sizable impact on other fields of science. It is the aim of this chapter to address this question. As there is a great variety of connections with other fields, we confine the discussions to a few illustrative examples. The chapter will be divided into four major parts, covering rather different aspects. The first section focuses on the cluster as a finite fermion system and compares cluster properties with similar finite fermion systems. In the second part, Section 6.2, we consider the role of clusters in astrophysics, while Section 6.3 focuses on the impact of clusters in climate. The last section, Section 6.4 finally presents several examples of applications of clusters in a biological context, especially in medicine.

6.1
Clusters in the Family of Finite Fermion Systems

The basic building blocks of matter are finite fermion systems: nuclei, atoms, molecules. Clusters, being a special subgroup of molecules, belong to this class. Finite fermion systems are built from a finite number of constituents whose properties are dominated by fermions and their particular quantum shell effects. In this section, we will discuss the place of clusters amongst the various finite fermion systems.

As presented in the introductory chapter, clusters are special in that they fill the range from a simple molecule all the way up to bulk matter by repeating the elementary building blocks in the proportions of bulk. Particularly special are metal clusters where the density of valence electrons has a tendency to keep close to a given (homogeneous) equilibrium density. We call systems with such a property "saturating" fermion systems. Table 6.1 summarizes and compares properties of saturating fermion systems. The column "Fermion" in the upper block shows

An Introduction to Cluster Science, First Edition. Phuong Mai Dinh, Paul-Gerhard Reinhard, and Eric Suraud.
© 2014 WILEY-VCH Verlag GmbH & Co. KGaA. Published 2014 by WILEY-VCH Verlag GmbH & Co. KGaA.

the quantum mechanically active constituent. Note that the three electronic systems carry different aspects. The ^{3}He droplets take the whole atom as the relevant quantum particle and the electronic properties summarized in the effective atom–atom potentials. Metal clusters also have the ionic cores as classically treated constituents which participate in the formation of structure and dynamical evolution. For quantum dots, on the other hand, one only handles the electrons and considers the substrate on which the dots are built as an inert background. The column "Size" provides typical constituent numbers. The last column indicates the corresponding "Bulk" material. The lower block shows typical length, momentum, and energy scales of the systems. All these saturating systems tend to maintain bulk equilibrium density. A consequence is the scalability of a system. For example, the density in the volume is almost independent of system size N, such that the radius scales with $N^{1/3}$. The ideal scalable systems are clusters which, indeed, have been produced at any size [12, 14]. That also includes droplets of ^{3}He, which are special clusters, distinguished by the fact that the whole atoms represent here the active fermions and that the bulk limit ends up in a Fermi liquid rather than in a solid. Nuclear matter is often considered as the bulk limit of nuclei. That, however, has to be taken with care. The steadily increasing Coulomb energy sets an upper limit of nuclear stability [150, 151]. Nuclear matter, although very useful for characterization, is a theoretical construction from which the Coulomb force has been removed to allow the bulk limit. Pure neutron matter, on the other hand, exists in principle at any size and is realized in neutron stars [152]. These are bound by gravitation which, being a long-range force, does not generate a universal equilibrium density (saturation). Quantum dots differ from the above three examples in that their size is defined by experimental construction. The quantum dots tend to constant electron densities if the same carrier material is used for the different sizes while the trapped clouds can be squeezed and deliberately expanded by tuning the external fields.

The lower block of Table 6.1 collects the typical length, momentum, and energy scales. Corresponding to saturation density ρ_{equil}, there exists a characteristic length scale $r_s \propto \rho_{\text{equil}}^{-1/3}$, usually known as the Wigner–Seitz radius. It provides the scaling of the radius R as $R \sim r_s N^{1/3}$. The Wigner–Seitz radius also induces a typical momentum scale in terms of the Fermi momentum, $\hbar k_F \propto r_s^{-1}$, which is the momentum of the least bound fermion, and subsequently, the Fermi energy $\varepsilon_F = \hbar^2 k_F^2/(2m)$ which is the corresponding kinetic energy. The latter thus sets a reference value for the energies in the system. From the Fermi momentum, one can estimate the de Broglie wavelength $\lambda_B \sim 2\pi/k_F \sim \pi r_s$. It spreads over about two interparticle distances, which means that the typical fermion wave function embraces several tens of constituents within its quantum mechanical width. This confirms the essentially quantum mechanical nature of finite fermion systems.

It is interesting to note that the systems collected in Table 6.1 span a huge range of scales, ranging from very small (fm) and strongly bound (MeV) nuclei to large (nm) and loosely bound (sub meV) helium droplets. Nonetheless, these systems have similar properties when expressed in their natural units as listed in Table 6.1.

Table 6.1 Schematic overview over saturating many-fermion systems. The first block of rows indicates typical sizes N, constituents, and the bulk limit. The second block of rows complements that by information on typical length, momentum, and energy scales.

Saturating finite fermion systems

System	Fermion	Size(N)	Bulk
Nucleus	Nucleon	$N \leq$ about 300	Limited by Coulomb
^{3}He droplet	^{3}He	$30 < N$	Quantum liquid
Cluster	Electron	$3 \leq N \leq 10^{5-7}$	Corresponding solid
Quantum dots	Electron	Defined by construction	Limited by construction

Scales of length, momentum, and energy

System	r_s	k_F	ε_F	Interaction
Nucleus	1.2 fm	1.35 fm^{-1}	35 MeV	Nuclear strong
^{3}He droplet	12 a_0	0.16 a_0^{-1}	2.7 K	Atom–atom
Metal cluster	3–5 a_0	0.4–0.6 a_0^{-1}	2–5 eV	Coulomb
Quantum dot	$\approx 200\, a_0$	$\approx 0.01\, a_0^{-1}$	1.4 meV	Coulomb

This is illustrated in Figure 6.1a, which shows the binding energy as a function of density for homogeneous matter of three different fermion fluids: nuclear matter, electron gas (representing the situation in simple metals), and liquid ^{3}He. It is remarkable that the three systems, which have dramatically different scales, all fit into one plot once energy and density are drawn in terms of the natural units (energy in ε_F, density in equilibrium density $\rho_{\text{equil}} = 3/(4\pi r_s^3)$). All binding curves have,

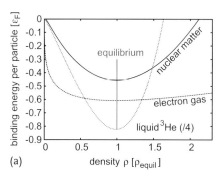
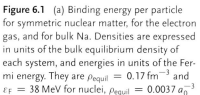
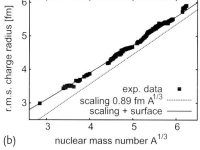

Figure 6.1 (a) Binding energy per particle for symmetric nuclear matter, for the electron gas, and for bulk Na. Densities are expressed in units of the bulk equilibrium density of each system, and energies in units of the Fermi energy. They are $\rho_{\text{equil}} = 0.17\,\text{fm}^{-3}$ and $\varepsilon_F = 38\,\text{MeV}$ for nuclei, $\rho_{\text{equil}} = 0.0037\, a_0^{-3}$ and $\varepsilon_F = 3.2\,\text{eV}$ for Na, and approximately the same for the electron gas. (b) Trend of nuclear radii with third root of nuclear mass number $A^{1/3}$. Data points are shown as boxes. The straight dotted line indicates a strict scaling $\propto A^{1/3}$. The solid line stands for a surface corrected trend.

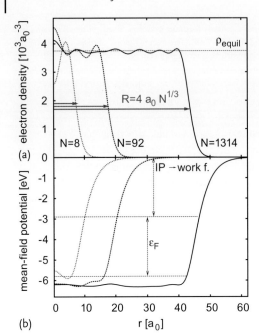

Figure 6.2 Electron densities (a) and mean-field potentials (b) for a Na cluster of different sizes as indicated. Fermi energy (ϵ_F) and ionization potential (IP) are nearly independent of the system, as indicated. Radii grow $\propto N^{1/3}$.

by definition, their equilibrium point at the same place. The scaled equilibrium energies vary only within a factor of two. Figure 6.1b demonstrates the consequence of saturation for the case of nuclear root-mean-square (r.m.s.) radii. The straight line indicates the estimate $R = r_s N^{1/3}$. The actual radii are close to this line but deviate with some small curvature. This can be explained as an effect of a finite surface width, which is the next correction beyond the simple bulk-scaling picture.

Figure 6.2 demonstrates the scaling properties in terms of detailed density and potential profiles for a series of Na clusters. Figure 6.2a shows how the density in the interior of the cluster nicely gathers around the bulk equilibrium values ρ_{equil} (indicated by a faint horizontal line). There always remains some small oscillations about ρ_{equil} with a wavelength of order of the k_F^{-1}. This is a quantum effect due to shell filling in discrete steps. The half-density radius (unlike the r.m.s. radius of the previous figure) perfectly increases $\propto N^{1/3}$. Figure 6.2b shows the corresponding potentials. The binding depth is also approximately constant and related to the Fermi energy ϵ_F plus bulk work function.

Shell effects are also crucial for example in determining the deformation of a finite fermion system. This was discussed in Section 4.1.2 for the case of metal clusters. Nuclei are showing a pronounced sequence of deformations as well, where shell closures (magic numbers) are associated with a spherical shape, while deformed ground states are typically found far off magic numbers [153]. Deformation is also the entrance door to fission which is a major reaction channel in heavy nuclei and whose properties are strongly determined by shell effects [154]. Accordingly, fission is also a conceivable process in metal clusters [21], although not reaching the importance it has in nuclear physics.

We have seen in Figure 6.2 one quantum effect, namely the shell oscillations in the density distribution. The most prominent shell effect is the appearance of magic numbers associated with filling a (spherical) quantum shell. We know that from atoms where noble gases are distinguished by particular strong binding and correspondingly large electron separation energy (IP). Magic numbers are illustrated in Figure 6.3 for four typical fermion systems. Except for nuclei in which fermions are neutrons and protons, the three other cases correspond to electronic shell closures. The magic numbers, of course, vary from one system to the next, due to different confining potentials. In the case of atoms (Figure 6.3a), shell closures can be directly read from the pronounced maxima of the ionization energies of neutral atoms for atomic numbers $Z = 2, 10, 18, \ldots$, which correspond to noble gases He, Ne, Ar, Kr, Xe, and Rn. The case of nuclei is illustrated in Figure 6.3c. The steps in separation energy (equivalent to electronic IP) reflect proton shell closures (2, 8, 20,

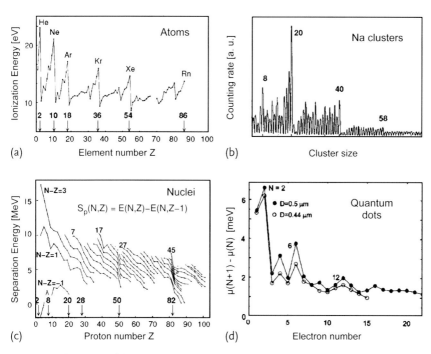

Figure 6.3 Illustration of the quantum shell effect "magic numbers" in finite fermion systems: electron separation energies (= IP) in atoms (a); abundance spectra of Na clusters (counting rate in arbitrary units) (b); proton separation energies in nuclei (connected lines have constant N–Z which is increased in steps of 2) (c); differences in the chemical potential (corresponding to IP) of disk-shaped quantum dots for two diameters D as indicated (d). Adapted from [155].

28, ...) in a way very similar to atoms. Figure 6.3b shows the case of Na clusters, this time considering as an observable the abundance of a species as a function of cluster size N. Shell closure corresponds to higher energy/stability thus larger abundance. One indeed observes pronounced maxima for clusters with $N = 2$, 8, 20, 40, and 58 atoms. Figure 6.3d shows the difference of chemical potentials (equivalent to IP) for the case of quantum dots. They have a planar geometry with a radial confining potential of harmonic oscillator shape. This produces the sequence of shell closures 2, 6, 12, ..., as is appropriate for a 2D oscillator.

Saturating fermion systems have also a prominent common feature in their excitation spectrum, namely pronounced collective resonances. We have already encountered them for metal clusters as the Mie plasmon resonance, see Sections 2.2.3 and 4.2. A similar feature exists in nuclei known as giant resonances [156, 157]. Figure 6.4 compares these two resonance modes. Figures 6.4b,d show the dipole strengths as a function of excitation energy. Both systems display very similarly one strong resonance peak. The peaks are well confined in energy. Remember that we have cut the plot to a very narrow energy window; there is however practically no strength outside. Figures 6.4a,c illustrate collective mechanisms beyond these modes. For the cluster (Figure 6.4c), the Mie resonance can be viewed as oscillations of the whole electron cloud against the frozen ionic background. This picture leads to the Mie estimate (2.6) for the resonance frequency and even allows for including next order refinements stemming from surface effects [30]. In the nuclear case (Figure 6.4a), one views the giant dipole resonance as oscillations of the proton cloud against the neutron one (called the Goldhaber–Teller mode). Its gross

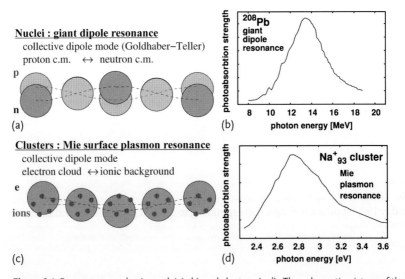

Figure 6.4 Resonance modes in nuclei (a,b) and clusters (c,d). The schematic picture of the collective motion underlying the resonance modes is given (a,c), as well as the optical absorption spectra (b,d).

properties can rather easily be deduced from simple sum rule considerations [158], much similar to the Mie frequency in the case of clusters (see Section 2.2.3).

6.2
Clusters in Astrophysics

Clusters can come into play in an astrophysical context since the composition of cosmic dust contains a major fraction of grains of ~ 100 nm size, the typical size of large clusters. The origin and the composition of these grains are various and often still under debate. Among many candidates, supernovae are possible contributors to cosmic dust. They are also the source of the heavy chemical elements. We particularly consider here the example of a supernova of type II (more than 8 solar masses $M_\odot$) which undergoes a gravitational core collapse, giving rise to a neutron star at its very center. The collapse ends up when the maximal neutron density is reached (due to Pauli exclusion principle). Then the external shells bounce and a shock wave propagates towards the outer layers with a velocity of about 10–20% of the speed of light. Energetic neutrinos, electrons and photons are emitted, and an active chemistry takes place: small molecules and clusters are formed from a gas phase, and can subsequently condense into dust grains. The chemical nature of the grains strongly depends on the initial composition of the ejecta. For example, an oxygen-rich environment produces more metal oxide and silicate dust, while SiC grains and amorphous carbon structures rather appear from a carbon-rich gas.

The theoretical description of such a chemistry represents a very complex task. For instance, in [159], as many as 159 chemical (bimolecular and radiative association) reactions implying He, C, O, Mg, Al, Si, S, and Fe are considered. The densities of molecular and cluster species are deduced from rate equations, from day 100 to day 1000 after the supernova explosion. A fully mixed ejecta as well as a stratified one have been explored. We present in Table 6.2 the percentage of each chemical element, initially in atomic form, locked up in one type of molecule or cluster at day 1000, for the explosion of a 20 $M_\odot$ supernova modeled by an unmixed ejecta. The ejecta is split into 4 zones corresponding to growing distance to the center (zone 1 is the innermost shell, zone 4 the outmost one) and accordingly different compositions corresponding to the original various star layers. The formed molecules serve as seeds to clustering, as is visible for $(SiO_2)_{1-5}$ in Figure 6.5. One can observe that, in zone 2, even if the atomic Si abundance is stable up to day 150, it suddenly drops, corresponding to the fast growth of abundances of the $(SiO_2)_n$ clusters. Interestingly, all clusters for $n = 1$ to 4 are subsequently suppressed, while the abundance of the $(SiO_2)_5$ cluster quickly levels off. This behavior is generic and is also obtained in all zones with other cluster species. The $(SiO_2)_5$ cluster, which exhibits a ring structure, can then coagulate into silicate grains.

It is worth noting that carbon clusters, including chains from C_2 up to C_9 and the closed ring C_{10}, are produced only in the outmost layers, which are C-rich, but in very low proportions (about 10^{-15} and 10^{-13} $M_\odot$ in zones 3 and 4, respectively). This production is hampered by oxidation reactions since the ratio O/C in zones

Table 6.2 Proportions of molecules and molecular clusters for various chemical elements, with respect to the total amount of corresponding atoms, 1000 days after explosion, obtained by an unmixed and stratified ejecta model of a type II 20 $M_\odot$ supernova [57, 159]. Zone 1 is the innermost shell while zone 4 is the outermost one. The main composition and the distance to the center (in solar mass unit) are indicated. In zone 4, the proportions of the C_{10} ring and of the CO molecule, indicated by stars, strongly depend on the presence of He^+, that is 0% in a He-rich zone, and 95.3% and 0.14%, respectively, in a He-free zone. Adapted from [159].

Atom	Molecule or Cluster	Zone 1 Si/S/Fe-rich 2.4–3 $M_\odot$ (in %)	Zone 2 O/Mg/Si-rich 3–3.6 $M_\odot$ (in %)	Zone 3 O/C/Mg-rich 3.6–4.95 $M_\odot$ (in %)	Zone 4 He/C/O-rich 4.95–5.85 $M_\odot$ (in %)
Si	SiS	28.3	...	...	0
	$Si_{4,5}$	3.1	49.5	48.2	0
	$(SiO_2)_5$	0.12	49.5	48.2	0
O	SO	0.7	...	...	...
	O_2	...	60.9	27.6	...
	$(SiO_2)_5$	...	3.9	...	...
	CO	...	...	32.7	...
S	SiS	91.4	...	...	...
	$(FeS)_4$	8.7	...	...	...
	SO	...	99.2	...	...
Fe	$(FeS)_4$	44.0	...	...	...
	Fe_4	0.1	...	...	...
Mg	Mg_4	...	0.91	0.04	0
	$(MgO)_4$	...	0.73	0	0
C	CO	...	99.8	99.7	0–0.14*
	C_{10}	...	...	...	0–95.3*
Al	Al_2O_4	...	99.1	96.8	0

3 and 4 is greater than 1. Note also that one of the major destructive reactions of molecules are due to their interactions with He^+ ions. This is especially true in the case of carbon chains. Indeed, the more the ejecta contains He^+ ions, the less and the later C_{10} rings are formed (see right-most column of Table 6.2). For instance, without any He in the environment, all C atoms are locked up in C_{10} clusters after day 160, while in the presence of 40% of He, the abundance of C_{10} is suppressed by three orders of magnitude and the time formation is as late as day 920.

The role of C_{10} rings formed in supernovae explosions is important as they can constitute precursors of fullerenes and polycyclic aromatic hydrocarbon (PAH) molecules. The latter are of great importance, since they have for almost thirty years been associated with the so-called unidentified infrared bands. These bands were first observed in the late 1970s in emission features of many celestial objects in our galaxy and in extragalactic regions submitted to UV flux, such as the planetary neb-

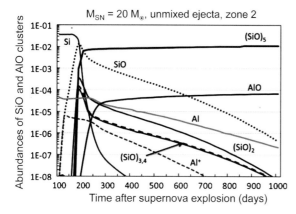

Figure 6.5 Time evolution of the abundance of SiO clusters and AlO molecules in the (O/Si/S-rich) zone 2, from an unmixed ejecta model for a $20 M_\odot$ supernova explosion. Adapted from [159].

ulae NGC 7027 [160]. The infrared measurements have made significant progress since then thanks to new space telescopes, such as the Spitzer Space Telescope (see for instance Figure 6.6a). However, the origin of the unidentified infrared bands remained a puzzle for almost ten years until the mid-1980s when small PAHs such as coronene ($C_{24}H_{12}$) and chrysene ($C_{18}H_{12}$) were evoked to produce many of the discrete bands, when irradiated by single UV photons in the laboratory. Numerous theoretical and experimental works on PAH have been dedicated to elucidate the whole mid-IR spectra of circumstellar and interstellar objects, see for instance [161] and references therein. Interestingly, in [162], the authors used complex statistical models to explore the IR emission of various carbonaceous sources (neutral and cationic PAH of various sizes, mixture of them, mixture with larger carbonaceous grains, heated by different starlight intensities, etc.) to deduce the interstellar dust mass or the PAH fraction from observed IR spectra. Figures 6.6c,d exemplify the case of PAH and PAH^+ molecules containing between 100 and 57 500 carbon atoms, and compares the IR emission to that of mixed PAH^+ and graphite spherical grains, whose diameters vary from 5 nm to 0.1 µm. Neutral PAH molecules exhibit stronger bands for wavelengths between 11 and 18 µm but also at 3.3 µm, while PAH^+ molecules dominate the emission in the 5–9 µm range. One can also notice in Figure 6.4d that the discrete bands tend to disappear with increasing PAH size, while a large band around 30 µm arises.

Note however that the hypothesis of PAH being the dominant carriers of the unidentified infrared bands is still a matter of debate. Indeed PAH alone are not sufficient to explain all the IR emission features (see [163] and references therein). In particular, it has been observed that some spectra exhibit a clear feature at 3.4 µm which cannot be accounted for by aromatic molecules but rather by aliphatic (that is which do not contain aromatic rings) compounds. The 6.8, 7.3 and 13.9 µm bands also seem to originate from long chain alkanes [164]. One can go even further and consider complex organic compounds consisting of amorphous networks of aro-

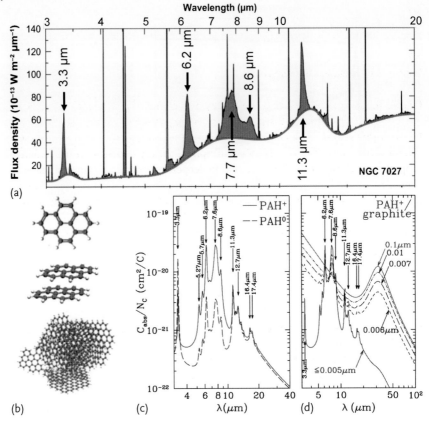

Figure 6.6 (a) Mid-infrared spectrum of the photo-dissociation region in the planetary nebulae NGC 7027. The "unidentified infrared bands" are indicated by arrows. (b) Examples of molecular structures, going from the pyrene molecule ($C_{16}H_{10}$), the pyrene dimer and a cluster consisting of pyrene molecules. Absorption cross-sections per C atom for neutral and ionized PAHs (c) which consist of 100 up to 57 500 C atoms (corresponding to size between 6 and 50 Å), and very small ionized carbonaceous spherical grains of indicated size (d). Composed from [162] and [161].

matic rings and aliphatic chains. But the variety of networks that one can possibly consider render the interpretation of the unidentified infrared band highly tricky. However a very recent article [165] aimed at demonstrating that, by analyzing in detail the relative intensities of the peaks at 3.4 and 6.8 μm in many IR emission spectra, the maximum fraction of C atoms locked up in aliphatic form should be less than 15%. In other words, the authors concluded that PAH indeed dominate in many cases the unidentified infrared bands. This encourages to further explore PAH spectroscopy in the gas phase to enlarge the available data and to decipher all the features observed in IR spectra from astronomical sources.

6.3
Clusters in Climate

6.3.1
Impact of Clusters in Climate Science

One of the major issues facing the twenty-first century is the global warming of the Earth. A key ingredient is given by radiative forcing, which is the difference between radiant energy received by the Earth and energy radiated back to space. The radiative balance of the Earth's atmosphere greatly depends on the radiative properties of clouds. It is thus essential to deeper understand the mechanisms of cloud formation. Indeed, almost forty years ago, the first evidence of the modification of cloud albedo by pollution were reported [166]. As a striking example, one can visualize in Figure 6.7 the formation of low-level clouds off the coasts of Alaska, precisely above shipping lanes. This is due to atmospheric aerosol particles, that is submicron liquid droplets in the gaseous atmosphere, emitted by the fuel burning of vessels while they travel over the sea.

Two complementary activities in aerosol science have attracted important efforts over the past two decades. On the one hand, such atmospheric particles have to be carefully detected in terms of size, concentration, chemical composition, optical properties, and so on. It is worth noting that the Scottish meteorologist J. Aitken built the first apparatus allowing the measurement of dust and fog particles in the atmosphere at the end of the nineteenth century [168]. Since then, several countries around the world have set dedicated geographical sites with sophisticated instruments for the measurement of aerosol sizes and concentrations, during specific campaigns or in a continuous manner. For instance, the European Supersites for Atmospheric Aerosol Research (EUSAAR) project has gathered since 2009 twenty

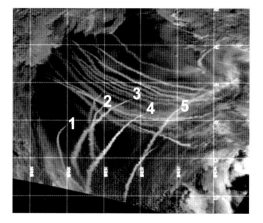

Figure 6.7 Five shipping lanes observed in the 2.3-μm images from the Moderate Resolution Imaging Spectroradiometer on the Aqua satellite off the coasts of Alaska in the early afternoon on 4 March 2009. The five corresponding plumes are indicated by numbers. Adapted from [167].

European sites, aiming at homogenizing the improvements of the instruments, the measurement protocols and the data delivery, in the same spirit as what was established in 2008 through the German Ultrafine Aerosol Network (GUAN), see [169] and references therein. Moving platforms such as aircrafts or ships are also used in such observational studies. For an exhaustive review, see for example [170] and the references therein, where more than 120 independent publications have been collected. Two main apparatus are used: the first category consists of a differential mobility particle sizer (DMPS), which measures sizes from 3 to 700 nm, while a condensation nucleus counter (CNC) gives access to concentrations. Note that in general, the chemical composition of aerosols is scarcely measured.

6.3.2
From Aerosols to Water Droplets

There is obviously an urgent need for detailed modeling. This is all the more true since the present limit in detection size is 3 nm (while the concentration resolution is at best 50 particles/cm^3) [171]. Aerosols are usually classified according to their diameter size, as summarized in Figure 6.8. One starts with ultrafine aerosols (or particles in the ultrafine mode) with a size smaller than 20–30 nm. In our language, they correspond to molecular clusters. This range of sizes is also denoted as the "nucleation" mode since it is basically this physical process, corresponding to the creation of a liquid droplet in a saturated vapor, which comes into play (see Section 6.3.3). Another contribution to this mode is the direct emission of such clusters from either natural sources such as sea-salt aerosols, or artificial ones such as combustion sources or industrial plumes. The ultrafine aerosols appear to be precursors to the so-called cloud condensation nuclei (CCN). The CCN are the entities which are studied the most in the aerosol field. Indeed they reach the size (around 100 nm) allowing for activation, that is coating by water. They are thus precursors of a cloud droplet. The corresponding Aitken mode will be detailed in Section 6.3.4.

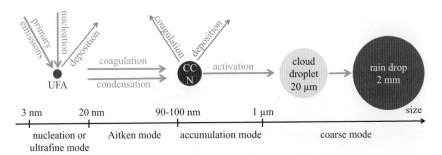

Figure 6.8 Schematic view of various types of particles relevant to cloud formation, from ultrafine aerosols (UFA) to cloud condensation nuclei (CCN), cloud droplets and finally rain drops, with typical diameter sizes. The mechanisms of formation and scavenging for each type of particle are indicated by arrows. See text for details. Compiled from [171, 172].

6.3.3
Formation Mechanisms of Aerosols

Coming back to the nucleation mode, different microscopic mechanisms have been proposed [171]: (i) binary water-sulfur nucleation, mostly in industrial plumes and free troposphere (closest air layer to the globe surface, between 8 and 15 km); (ii) ternary water-sulfur-ammonia nucleation, mostly in the continental boundary layer (air layer near the ground affected by diurnal heat, moisture or momentum transfer to or from the surface); (iii) ion-induced nucleation, in upper troposphere and lower stratosphere; (iv) organic-involved nucleation, in the same areas as in (iii); (v) barrier-less homogeneous nucleation of iodide species, essentially in coastal environments. To illustrate the correlation between the concentration of sulfuric acid and the formation of ultrafine particles, we present in Figure 6.9 their concentration as a function of time (from 0.00 a.m. to 12.00 p.m.), recorded during a nucleation event on 21 September 1993 at Idaho Hill, Colorado [173]. The concentration of H_2SO_4 abruptly increases after sunrise and peaks around 10 a.m. It then decreases and vanishes a bit after sunset. Interestingly, there is a clear correlation with the ultrafine aerosol concentration pattern, which looks very similar to that of H_2SO_4 but delayed by 1.5 hours. These measurements corroborate the nucleation mechanism (i) mentioned above.

Other nucleation processes are explored as, for example, mechanism (iii), that is ionic nucleation, since high-energy cosmic rays can form ions in the lower atmosphere. For instance, protonated ammonia-water clusters $H^+(NH_3)_m(H_2O)_n$ with $m = 4-6$ and $n = 1-25$ are formed in the laboratory by corona discharge in ambient air [174]. Magic numbers are experimentally observed (see Figure 6.10a), as $H^+(NH_3)_5(H_2O)_{20}$, with one of its optimized structures calculated with a semiempirical Hamiltonian implemented in the MOPAC (Molecular Orbital PACkage) code [175] depicted in Figure 6.10b. Interestingly, the high stability of protonated ammonia-water clusters containing five NH_3 molecules does not seem to rely on the water-coating of the very stable tetrahedral structure of four equivalent

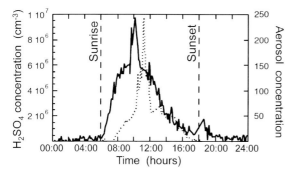

Figure 6.9 H_2SO_4 (solid curve, left vertical scale) and 2.7–4 nm diameter ultrafine aerosols (dotted, right vertical scale) concentrations as a function of time, measured at a remote continental site located at Idaho Hill, Colorado, during a nucleation event on 21 September 1993 [173]. Adapted from [170].

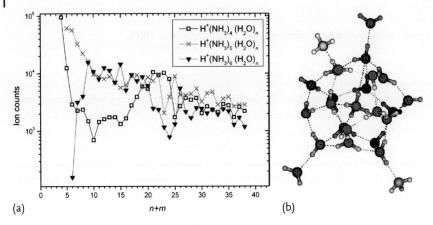

Figure 6.10 (a) Ion counts of produced $H^+(NH_3)_m(H_2O)_n$ as a function of $n + m$ in a discharge experiment. (b) Example of an optimized structure for $m = 5$ and $n = 20$, calculated with the MOPAC code [175]. See text for further details. Adapted from [174].

NH_3 linked to a NH_4^+ core, but rather on a competition between H_2O and NH_3 molecules around the central NH_4^+. This kind of mixed cluster can therefore act as nucleation seeds in ammonia-containing water vapor.

Once the ultrafine particles are created, mostly by (homogeneous or heterogeneous) nucleation, their size can grow by self-coagulation or by condensation of low-volatility gases as H_2SO_4 again, HNO_3 or NH_3 (see discussion above), onto the cluster surface. Note that one often distinguishes mechanisms of nucleation and those of growth, since the chemical conditions in the atmosphere might be different. More generally, the observed nucleation rates of 3 nm diameter particles are 0.01–$10\,\text{cm}^{-3}\,\text{s}^{-1}$ in the boundary layer (air layer near the ground), greater than $100\,\text{cm}^{-3}\,\text{s}^{-1}$ in urban areas, and 10^4–$10^5\,\text{cm}^{-3}\,\text{s}^{-1}$ over coastal zones and in sulfur-laden industrial plumes [170]. Typical growth rates lie in the 1–20 nm/h range; however in coastal areas, they can be as high as 200 nm/h, compared to the very low 0.1 nm/h growth rate in clean polar areas. Moreover, nucleation and/or growth rates can also depend on the season. For instance, the aerosol concentration is higher during summer than during winter in the Alps, while at the EUSAAR Italy site, concentrations are the largest in winter.

6.3.4
Clusters as Seeds for Cloud Condensation Nuclei

The distinction between ultrafine aerosols and cloud condensation nuclei (CCN) does not refer only to their diameter, see Figure 6.8. As mentioned above, the specificity of a CCN is its ability to uptake water molecules. This depends on some external conditions as the relative humidity (RH) of the air around the CCN. More precisely, if S is the relative saturation ratio, that is the partial water pressure p over the saturation vapor pressure P^* at the same temperature, we define RH =

$p/P^* \times 100\% = S \times 100\%$. We usually have RH $< 100\%$. But if the vapor is supersaturated, that is when $p > P^*$, we have RH $> 100\%$.

One can show that, at a given S, the variation of Gibbs free energy during the formation of a spherical water liquid droplet of radius r within a vapor phase is equal to $\Delta G = -4\pi r^3 k_B T \ln S/(3v_l) + 4\pi\sigma r^2$, where v_l is the volume of a water molecule in the liquid phase, k_B the Boltzmann constant, T the temperature, and σ the surface tension of the water/air interface [176]. If $S \leq 1$, then $\Delta G > 0$ and the formation of liquid droplet is not thermodynamically favored, whatever r. On the contrary, if $S > 1$, that is when the vapor is supersaturated, the two terms in ΔG have opposite signs and there thus exists a maximum at a critical radius r_c given by $r_c = (2\sigma v_l)/(k_B T \ln S)$. Therefore, when $r > r_c$, a liquid water droplet can form. It is also thermodynamically favorable that this droplet grows and r increases. This relation between S and r_c is better known as the Kelvin equation if one expresses S as the ratio of the saturation vapor pressure above a curved liquid water surface with r_c as the curvature radius, over that of a flat ($r \to \infty$) water surface:

$$S_{\text{Kelvin}} = \frac{P^*(r_c)}{P^*(\infty)} = \exp\left(\frac{A}{r_c}\right) > 1, \tag{6.1}$$

where $A = 2\sigma v_l/(RT)$, where R is the ideal gas constant. The fact that this ratio is greater than 1 is coined the Kelvin effect, which states that the saturation pressure over a curved surface is always greater than that over a flat surface. This can be intuitively understood since molecules at a surface share fewer neighbors if the surface is curved than if it is planar, leading to fewer interactions and thus more evaporation.

There is however a countereffect to the Kelvin one, which tends to reduce the radius of a growing liquid droplet, if the water droplet heterogeneously forms. This is due to Raoult's law: consider a solution consisting of n_w moles of water molecules and n_s moles of solute molecules. In the case of flat interfaces, it states that the partial vapor pressure above the solution divided by the saturation vapor pressure at the same temperature reads $p(\infty)/P^*(\infty) = n_w/(n_w + n_s)$. This ratio is also called "water activity" and denoted by a_w. For a surface of curvature r, it reads as:

$$a_w = \frac{p(r)}{P^*(r)} = \exp\left(-\frac{B}{r^3}\right), \tag{6.2}$$

where B is constant depending on the mass of the solute molecule, that of the water molecule, and the water mass density. This version of Raoult's law shows that dissolving a solute in a liquid water droplet of radius r prevents water molecules from evaporating, or in other words, decreases the saturation vapor pressure. Inversely, if the solution droplet grows, it gets more dilute: the Raoult effect is less effective and the droplet tends to evaporate water molecules.

If we now combine Eqs. (6.1) and (6.2), one obtains the saturation ratio of the Köhler theory [177]:

$$S(r) = \frac{p(r)}{P^*(\infty)} = a_w \cdot S_{\text{Kelvin}} = \exp\left(-\frac{B}{r^3} + \frac{A}{r}\right). \tag{6.3}$$

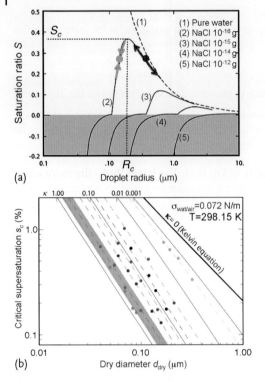

Figure 6.11 (a) Köhler curves according to Eq. (6.3), in the case of water droplet containing small amounts of solid NaCl. The dashed curve, labeled 1, corresponds to the Kelvin equation (6.1). The critical saturation S_c and the critical radius R_c are indicated for curve (2). See text for details. Adapted from [176]. (b) Critical supersaturation s_c as a function of dry diameter d_{dry} in the κ-Köhler theory, see Eq. (6.4). Calculations with temperature of 298.15 K and water/air surface tension $\sigma = 0.072$ N/m, are plotted as lines for different values of κ (see upper scale). Observed supersaturations reported for various types of solute molecules (organic, inorganic, mix of both) are displayed as circles. Results from each inserted solute line up on almost straight lines (see dashed fitted lines), according to (6.4). Adapted from [178].

For some given solute molecules, the plot of S as a function of r represents the Köhler curve. Examples are drawn in Figure 6.11a for the case of a water droplet containing small amounts of solid NaCl. It exhibits a maximum at a critical radius R_c (not to be confused with the above r_c in the Kelvin effect of (6.1)). The corresponding value of the saturation is called the critical saturation ratio S_c. If $r < R_c$, see the light circle on curve (2) in Figure 6.11a, the equilibrium is stable since a small change in r and/or in S leads either to condensation or evaporation so that, the original droplet radius is recovered. Inversely, if $r > R_c$, and if $S_c > 1$, which is the case of the dark circle on curve (2), the equilibrium of a water droplet containing solute molecules is unstable: a tiny increase of r produces a Kelvin effect which overcomes more and more Raoult's law. The droplet spontaneously grows – it is then activated. In other words, a nanoparticle with a radius greater than R_c acts as a cloud condensation nucleus (CCN).

This theory is however not applicable as such to nanometer droplets containing multiple solutes. It has thus been extended in the past decade, for instance as the κ-Köhler theory [178]. In the latter, the water activation a_w is taken as a function of the radius r of the wet nanoparticle (the CCN), that of the dry nanoparticle r_{dry} (note that in the Köhler theory, solute molecules are needed to produce Raoult's law), and a parameter κ called the hygroscopicity. More precisely, the saturation ratio is written as:

$$S(r) = \frac{r^3 - r_{dry}^3}{r^3 - (1-\kappa)r_{dry}^3} \exp\left(\frac{A}{r}\right), \quad (6.4)$$

where the constant A is the same as in Eq. (6.1). The larger the κ, the more active the CCN. As in the Köhler theory, S in the κ version also exhibits a maximum for a given r_{dry}. By solving $(\partial S/\partial r)|_{r_{dry}} = 0$, one determines the radius for which S reaches its maximum, namely the critical saturation ratio S_c, which in turn is a function of r_{dry}. S_c is interpreted as the saturation ratio needed for a nanoparticle of a certain dry size to act as a CCN. If one furthermore neglects the solute volume, the relation between S_c and r_{dry} simplifies to the following form $54\kappa r_{dry}^3 \ln^2 S_c = A^3$. Figure 6.11b plots the critical supersaturation, defined as s_c (%) $= 100\times(S_c-1)$, as a function of the dry diameter of the nanoparticle d_{dry}, for different values of κ [178]. Experimental data are also displayed, corresponding to different chemical composition of the CCN (inorganic, organic, various mix of both, …). For a given compound, agreement with the κ-Köhler theory is very good, since the observed data nicely lie on the dashed lines. This allows one to deduce the corresponding value of κ. Inorganic compounds exhibit κ close to 1 (see shaded area); the largest observed value is $\kappa = 1.3$ for NaCl (not shown here). On the contrary, organic molecules have CCN activities corresponding to $0.002 < \kappa < 0.5$. Note that one recovers the Kelvin equation (6.1) in the limit $\kappa = 0$. The κ-Köhler curves also show that the smaller the dry diameter, the larger the needed supersaturation to activate the aerosol particle and to qualify it as a CCN. The power of the κ-Köhler theory relies on the use of the single parameter κ to describe the water activity of CCN which can consist of complex chemical composition (single or multicomponent solution).

6.4
Clusters in Biological Systems

Clusters in biology and medicine constitute a huge developing field, see for example [179–181]. A large variety of nanoparticles (gold clusters, fullerenes, superparamagnetic clusters, …) have been designed to perform imaging, diagnosis, or therapy, sometimes simultaneously. However one major difficulty of such applications concerns an efficient functionalization of the cluster, that is the modification of its surface. Complex chemical procedures come into play here. The aim is twofold: to obtain biocompatibility, and to correctly position the cluster on the target biosystem (cell, unit in a cell, protein, etc.). This also implies *in vitro* and *in vivo* studies of possible toxicities of such clusters: this is the development of the field of nanotoxicity in general.

Once the functionalization and the nanotoxicity of a cluster is (more or less) under control, one exploits specific physical properties of the cluster to achieve a given medical goal, for instance a significant increase of the sensitivity, the resolution, and/or the stability in time of imaging techniques. It is worth noting that, in most of the examples presented in this section, it is the preferred coupling of a cluster to light, especially when delivered by a laser pulse, which is used in medical applications. For instance, one can tune the laser frequency to resonantly excite a metal cluster through its surface plasmon resonance (see Sections 2.2.3 and 4.2) and enhance electronic emission (see Section 4.3). In other words, the metal cluster acts here as a chromophore compared to the molecules around. Indeed, the Mie plasmon allows the metal cluster to absorb more photons at a given energy, typically in the visible domain, whereas water (the dominant constituent of the body) is transparent in that frequency range. Note that the notion of a chromophore does not only refer to metal clusters with well-defined plasmon resonances. It can appear in a more general context, especially in biology where fluorescent organic dyes are indeed denoted as chromophores. But the underlying mechanism remains the same: some delocalization of valence electrons can give rise to collective modes which allow a resonant coupling with an external electromagnetic field. This is indeed the kind of resonance which is exploited in medicine, as will be illustrated in Sections 6.4.2 and 6.4.3.

We shall by no means be exhaustive here. Our aim is to give a few possible applications of clusters in biology and medicine, among many others. In particular, quantum dots (see Section 6.1) are widely used in this context. But to remain specific to clusters only, we do not address the use of quantum dots in medicine. This section is organized as follows. We first briefly give some examples of cluster dynamics in living tissues, implying a specific tailoring of clusters. We then review some of the numerous medical applications of clusters, namely imaging and diagnosis on the one hand, therapy and drug delivery on the other hand, by illustrating them with some striking examples. We finally end with the societal issue of nanotoxicity.

6.4.1
Tailoring Clusters

Depending on the means used to inject a cluster in a living tissue, it can encounter several obstacles on its way before reaching the targeted tissue. A basic issue here is metabolic clearance, namely the capability of the tissue to eliminate the cluster. But the stability or some chemical or physical property of the cluster can also be affected by the environment provided by a living tissue. For instance, iron clusters would be ideal for high-resolution nuclear magnetic resonance imaging thanks to their high magnetic moment (see Figure 2.6). However, aqueous environments, typical of biological systems, quickly oxidize them and convert them into nonmagnetic compounds. It is thus mandatory to encapsulate the clusters to prevent the

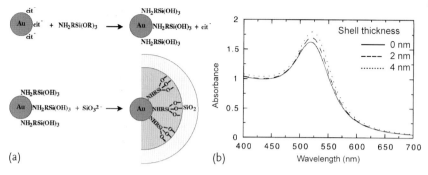

Figure 6.12 (a) Schematic representation of silica encapsulating of 15 nm-diameter gold clusters via reduction of citrate and a silane (here NH$_2$RSi(OH)$_3$, where R is an alkyl group). (b) Absorbance spectra of these gold clusters for different silica shell thicknesses in water solution. Adapted from [182].

loss of their magnetic properties. The dynamics of clusters in a biological environment should also be considered, since small clusters, more mobile than larger size clusters, have a tendency to aggregate. This can result, on the one hand, in a strong modification of physical properties of the clusters (as, for example, optical absorption, see Section 2.2.3), and, on the other hand, in the impossibility for the cluster to be taken up by a cell for instance. A proper modification of the cluster surface can prevent such an aggregation and allow a desired "solubility" in water. Sometimes, it can even enhance some physical property the cluster is used for, again such as optical response.

As an example of cluster tailoring, we mention the well-established citrate reduction techniques for the embedding of gold nanoparticles in silica shells. The citrates used here are salts of citric acid. One of them is presented in Figure 6.12a [182]. The idea here is to attach silica on the surface of the gold cluster which is a difficult issue since gold has very little affinity for silica. This is done in two steps. Indeed, gold clusters, which are formerly covered by citrate anions, can then form complexes with silanes (i.e., compounds containing at least a Si–OH group) by replacement of the citrates, giving rise to a monolayer of silanes on the cluster surface. The subsequent addition of a solution of sodium silicate provides active silica, which can coat the silane-covered gold clusters via bonds with the Si–OH group. The silica shells then prevent the aggregation of the gold clusters in water and also induce a slight enhancement of their plasmon peak intensity (cf. Figure 6.12b) with a very weak red-shift of the peak position.

Vascular targeting is also of great medical importance since tumor and normal vasculatures exhibit different characteristics, as for instance different flow conditions. There exist some stochastic models describing the adhesion and the absorption by cells of circulating spherical clusters decorated by ligands matching receptors present on the vessel walls [183]. This work reports a "design map" showing an optimal region where adhesion and absorption can happen at the same time, for given external conditions (density of receptors, size of the clusters, surface binding affinity, etc.). This corresponds to cases where the force between ligands and receptors is high enough to allow adhesion under the blood flow, and where the ratio of number of ligands over that of receptors is not too high, otherwise, the decorated nanoparticle would be too large to penetrate the vessel wall.

6.4.2
Clusters for Medical Imaging

In the previous section, we have presented some ways to encapsulate clusters to allow their use in a biological context. One usually combines them with some functionalization of the coated cluster, either to design a physical, chemical or biological sensor, or to mark a specific protein, a cell membrane, or a cell, with some possible alteration of the physical (typically optical) properties the cluster provides. In the following, we therefore distinguish sensors from markers in medical imaging.

6.4.2.1   **Markers**

Magnetic resonance imaging (MRI) is a commonly used technique in diagnosis imaging. It relies on the measurement of the precession of the nuclear spin of hydrogen atoms in water molecules, induced by an external magnetic field. To discriminate between one specific tissue and another, contrast agents attached to the target tissue are used to accelerate the spin relaxation of nearby water molecules. Superparamagnetic ion oxide nanoparticles (SPION) are explored for such imaging. For instance, in the case of neuronal stress or inadequate cerebral blood supply (ischemia), a specific peptide can be expressed one or two orders of magnitudes more than in normal cerebral activity. This is due to an abnormal fabrication and transcription of a specific RNA, which then represents the target RNA. An RNA is a single strand, analogous to a DNA single strand. We remind that DNA has a double-strand structure, where two opposite strands are denoted as complementary. Indeed the four DNA bases, namely adenine, thymine, cytosine, and guanine, do not randomly pair: adenine only binds to thymine, while cytosine binds to guanine. In RNA, thymine is replaced by uracile, and thus pairs up analogously as DNA does. Therefore, if one wants to detect a given RNA (the target), one uses the complementary RNA as a marker. Coming back to our example, one thus attaches the complementary RNA on SPIONs whose detection by MRI will indicate whether a dysfunctional transcription of the target RNA occurs or not. This technique has been successfully applied in living animals and also *in vitro* [184].

Another type of MRI contrast agents uses noble metal clusters, such as Au_3Cu_1 hollow nanospheres, covered by biocompatible polymers [185]. A TEM image of such nanosystems is presented in Figure 6.13a where the polymer shell is clearly visible. *In vitro* experiments demonstrated that the magnetic resonance signal of water monotonically increased with the concentration of nanospheres in water, already with a concentration as low as 0.125 mg/mL. The signal enhancement factor was 89% for a nanosphere concentration of 5 mg/mL. The enhancement mechanism probably relies on the porous hollow morphology of the nanosphere, thus creating specific interactions with water molecules. From the medical point of view, these nanospheres thus constituted potential contrast agents of blood vessels.

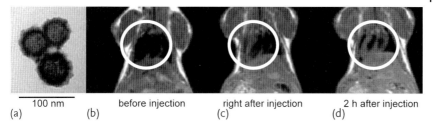

Figure 6.13 (a) TEM image of contrast agents polymer-coated Au_3Cu_1 nanospheres. Shown is a sequence of nuclear magnetic resonance images of the cardiac region (longitudinal view) of mice where contrast agents have been injected in the tail vein, to image the mice's heart (inside the circles): before (b), right after (c) and 2 h after injection (d). Adapted from [185].

In vivo tests have been carried out on mice to check their MRI efficiency in highly vascularized areas. In Figure 6.13, MRI images of the cardiac region of mice have been recorded before injection of the nanospheres (Figure 6.13b), right after (Figure 6.13c), and 2 h after injection (Figure 6.13d). The aim here was to visualize the mice's heart (inside the indicated circles). A higher contrast was clearly achieved right after injection of the polymer-coated Au_3Cu_1 nanospheres, and the contrast even increased 2 h later. Such nanosystems therefore constitute efficient blood-pool agents.

6.4.2.2 Sensors

In contrast with markers, a sensor is designed to exhibit different physical properties when it has either detected or not detected the target molecule. As discussed in the case of embedded clusters, see Section 5.1.2, the surface plasmon resonance of a metal cluster strongly depends on its direct environment, for instance when it is functionalized to design a sensor, and when the sensor itself binds to the target analyte.

This property has been used in a recent imaging technique which combines the surface plasmon of gold nanoparticles (GNP) and the absorbance of a protein in the so-called plasmonic resonance energy transfer (PRET) method [186]. The idea is to match the optical response of a metal cluster with absorption peaks of a biomolecule. An energy transfer of the plasmon resonance to one or several resonant peaks of the biomolecule then becomes possible and results in a modified optical response of the GNP. Figures 6.14a,b schematically illustrate such a transfer, while Figure 6.14c demonstrates it in *in vitro* experiments [187]. Here, GNP are functionalized with a carboxyl acid which can serve as a weak ligand to a certain protein. The latter one is here a natural intracellular chromophore present in some cells, called the cytochrome *c*. We have here two different chromophores (or photosensitizers). First, the cytochrome *c* exhibits narrow absorption lines at 525 and 550 nm (see light gray curve in Figure 6.14a). Second, the GNP possess a plasmon resonance whose position can be chosen by a proper size of the GNP. And indeed, the GNP's diameter is designed so that its plasmon peak can overlap the absorption peaks of the cytochrome *c*, as sketched in Figure 6.14a. The mechanism of

PRET is then the following. One illuminates a GNP-coated by cytochrome c in the Rayleigh regime, that is with a laser of wavelength much larger than the GNP size. The Rayleigh scattering allows an excitation of the plasmon of the metal cluster (the donor). The latter one can be then absorbed by the cytochrome c (the acceptor) if the plasmon matches one of its two absorption lines. The resulting Rayleigh scattering consequently exhibits quenching dips, precisely at the absorption wavelengths of the cytochrome c, which then gives a direct signature of the presence of the cytochrome (see Figure6.14b). The experimental measurements are depicted in the Figure 6.14c. Two GNP sizes are presented, the larger one giving rise to a plasmon maximum at 590 nm, the other one at 550 nm. As expected, the PRET better operates when the plasmon peak of GNP coincides with one of the two absorption peaks of the cytochrome c, as demonstrated by the very deep quenching dip at precisely 550 nm, while the quenching dips in the larger GNP are much less pronounced. The same technique has also been successfully applied in *in vivo* experiments where the time evolution of the depth of quenching dips is recorded. This thus allows to monitor in real time concentrations fluctuations of a resonant

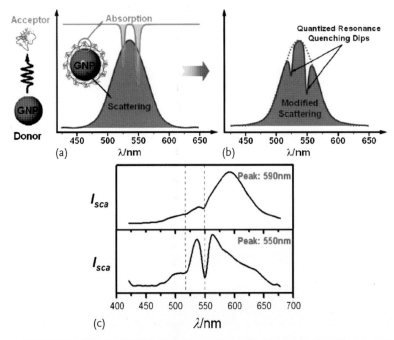

Figure 6.14 (a) Absorbance (light gray curve) of a protein naturally in cells (here, conjugate cytochrome c), with maxima at 525 and 550 nm, on top of the Rayleigh scattering (dark gray curve) of bare gold nanoparticles (GNP). (b) Schematic picture of modified Rayleigh scattering of a functionalized GNP bound to the protein. (c) *In vitro* measured Rayleigh scattering, in the presence of the protein, for a GNP exhibiting a plasmon peak at 590 nm, and for another one with a plasmon peak at 550 nm. The vertical dotted lines indicate the position of the absorption peaks of the protein. Adapted from [187].

bio-molecule (here the cytochrome *c*) in a living cell, thanks to the reversible binding of this chromophore to the functionalized GNP [187].

Single-wall carbon nanotubes (SWCNT) also represent versatile biosensors when their surface is properly functionalized, see for example [188] and references therein. Indeed such carbon nanotubes are well suited since their surface charge density is quite similar to that of proteins. Therefore, the electrostatic interactions between functionalized SWCNT and analyte biomolecules allow sensitive electrical detection. For instance, the phenylboronic acid, whose chemical structure is $C_6H_5B(OH)_2$, can be used as a glucose sensor. Such biomolecules can bind to SWCNT covalently or noncovalently, depending on the chemical process used there. The type of bonding significantly affects the sensing mechanism: when glucose attaches a phenylboronic acid covalently bound to a SWCNT, the resistance of the latter is increased, while in the case of a noncovalent binding, the SWCNT resistance is not affected but the gate voltage characteristics and the gating efficiency are modified.

SWCNT can also be functionalized to produce very sensitive DNA sensors [189]. Detection of tiny concentrations of DNA represents a promising tool for fast on-site diagnosis of various diseases, whose signature can often consist of emission of a few copies of DNA. To allow such a detection, a standard means in molecular biology is the so-called polymerase chain reaction (PCR). This technique generates thousands to millions of copies of a particular DNA sequence, and thus amplifies the DNA concentration by several orders of magnitude. However, it often takes several hours to realize a PCR, and it is usually very expensive. New techniques employing functionalized SWCNT prevent the use of PCR and offer the advantage of a compact and portable sensing device. The latter relies on an electrical measurement, thanks to the highly conducting properties of SWCNT mentioned above. To be more specific, an example of such a functionalized SWCNT is presented in Figure 6.15a, where probe single-strand DNA is fixed on its surface. The aim here is to detect a given single-strand DNA in solution. One thus attaches on the SWCNT the complementary DNA, so that only the target DNA can bind, similarly to the detection of a target RNA discussed above (see beginning of the paragraph on markers). The sensitivity of the sensor has been tested in two different solutions. The first one contained three different single-strand DNAs but none was the target DNA. Therefore they could not hybridize with the complementary DNA bound to the SWCNT. The DNA concentration was 9 fmol. The second solution was the same as the previous one but with the addition of 200 amol of the target DNA. The complex impedance Z of such devices was measured in a buffer solution as a function of the frequency response and the gate voltage, thus giving a reference measurement. Z was then measured in the two solutions described above. A shift in the gate voltage dependence, compared with the reference response, was observed in both cases. The shift was however enhanced by a factor 5 in the presence of the target DNA, even if the latter one had a concentration 45 times lower than the three noncomplementary DNA in the solution. This demonstrates the high sensitivity of such a complex sensor with DNA concentration.

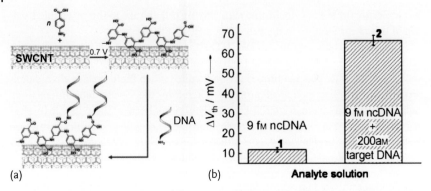

Figure 6.15 (a) Schematic view of single-wall carbon nanotubes (SWCNT) functionalized first with an amino-acid by electro-polymerization under a 0.7 V voltage, and second with a NH$_2$-single-strand DNA. (b) Sensor signal (that is, the shift of the measured impedance as a function of gate voltage, with respect to a reference measurement) in the case of no complementary DNA in solution (left box) and with a very small amount of complementary DNA in addition to the solution (right box). See text for details. Adapted from [189].

6.4.3
Clusters for Therapies

6.4.3.1 Drug Delivery

The amazing progress in nanoengineering allows one to dream about complex drug carriers which could be orally administrated, pass the wanted cell membranes, and deliver the drug to the targeted site in a time-controlled manner. The release of the drug basically relies on the breaking of bonds with the drug carrier. This can be controlled for example by the environment in tumor cells (as a specific pH or the presence of inherent biomolecules) or by the application of an external electromagnetic field. We present here some of the numerous solutions already explored clinically, see for example [190] for a recent review. Metal clusters in this respect are once again well exploited.

Gold nanoparticles (GNP) have for example been used to carry a radiosensitizer cancer agent to a tumor more rapidly than when the agent is injected alone [191]. More precisely, the silicon phthalocyanine 4 (Pc4) is a chromophore, which by absorbing light at 672 nm can produce a local heating sufficient to kill surrounding cells (cf. cancer photothermal therapy below). This molecule is however highly hydrophobic and, therefore, cannot be directly injected in the blood circulation. To circumvent this, one first exploits the weak binding of Pc4 onto a GNP surface. Second, the GNP–Pc4 complexes are dressed with polyethylene glycol (PEG) polymers to render them water-soluble and to prevent cluster aggregation. This coating also inhibits protein adsorption. This means that such a GNP–Pc4–PEG compound is not easily eliminated from blood by the body's immune system. The circulation time of the complex in blood is therefore much more extended compared with that of Pc4 alone. Last but not least, it has also been observed that this kind of drug vector preferentially accumulates in tumor sites, due to a less efficient vasculature there. *In vivo* tests in mice have demonstrated, whereas Pc4 usually takes 2 days after injection to reach the tumor, the PEG–GNP–Pc4 complex is almost instantaneously transported to the tumor (< 1 min) and is accumulated within 2 h. Functionalized gold clusters are therefore used here as an efficient passive drug carrier.

As a second example, we present a case where gold clusters are functionalized with single-strand DNA, which serves as a ligand to a specific cancer prodrug [192]. A prodrug is a biologically inactive compound that can be metabolized in the body to produce a drug. The one used in this experiment contains a Pt atom which binds to 2 Cl atoms, 2 NH_3 molecules, 1 OH group and 1 O atom. It is very close to the most commonly used antitumor drug cisplatin which consists of a Pt atom bound to 2 Cl atoms and 2 NH_3 molecules. Due to the different number of bonds, the Pt atom possesses in the prodrug (6) and in the cisplatin (4), the Pt atom has an oxidation level of IV in the prodrug, while it has a oxidation level of II in the cisplatin. This means that the cisplatin is electronically more active than the prodrug. This property actually allows the cisplatin to attach DNA in a cell and to kill the corresponding cell. In other words, the Pt(IV) atom of the prodrug turns out to be less toxic in cells than the Pt(II) atom of the cisplatin. In [192], the prodrug is chemically attached to a single-strand DNA, the latter one being linked to a GNP surface. The action of Pt(IV)–DNA–GNP complexes relies on a pH in tumor cells which is lower than the physiological one (5–6 instead of 7.4). This acidic pH in cancerous cells therefore allows for a reduction of the oxidation level of Pt from IV to II, and the release of cisplatin. *In vitro* experiments in different cancerous cells have been carried out and demonstrated an efficiency of the Pt(IV)–DNA–GNP conjugates as good as cisplatin, or even better in some cases. The pH-induced delivery of an anticancer drug from such functionalized gold clusters with an inactive prodrug thus represents a promising route for a chemotherapy with fewer side effects than the numerous ones provoked by cisplatin alone.

Drug release can also be controlled by light. For instance, some polymer hydrogels, priorly loaded with proteins, exhibit a reversible collapse of their structure when the local temperature is increased [193]. This geometrical modification induces in turn the release of the loaded proteins. The polymers' composition is chosen so that the structure collapse occurs at about 40 °C. To finely control the increase of temperature, 37 nm-diameter Au_2S clusters coated by 4 nm Au shells have been designed to efficiently absorb light in the so-called therapeutic window ranging from 700 to 1200 nm, while the polymer hydrogel hardly absorbs in this frequency range. This window corresponds to a dramatic drop of the photoabsorption in water and in human tissues [194]. An infrared laser pulse can therefore heat the gold composite nanoparticle and raise the polymer's temperature up to 40 °C to induce a contraction of its structure. For instance, the delivery of bovine serum albumin (BSA) loaded on this polymer hydrogel has been studied after irradiation by a nanosecond laser pulse of a frequency of 1064 nm, as illustrated in Figure 6.16a. At the end of the 40 min of irradiation, when gold composite clusters are used, one can obtain about twice the amount of BSA from noncomposite polymer hydrogels.

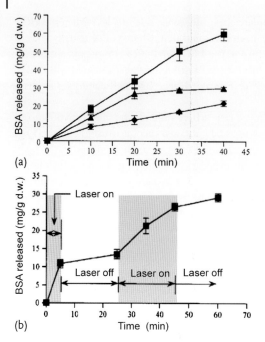

Figure 6.16 (a) As a function of time, bovine serum albumin (BSA) release from nonirradiated polymer hydrogels (diamonds), irradiated polymer hydrogels (triangles) and irradiated composite nanoparticle hydrogels (squares) during 40 min by a laser with the following characteristics: 1064 nm, 164 mJ/pulse, 7-ns pulse length, 10-Hz repetition rate. (b) Release of BSA from composite nanoparticle polymer hydrogels irradiated by the same laser but first during the first 5 min. and then during the period between 25 and 35 min. Adapted from [193].

This measurement clearly demonstrates the chromophore effect induced by the metal clusters. Moreover, since the collapse of the polymer hydrogel is reversible, when all proteins are not delivered during a first period of irradiation, sequential irradiation can allow the polymers to release the remaining molecules during a second stage, see Figure 6.16b. This kind of light-controlled drug release is very promising for pulsatile deliveries of drugs, as in insulin therapy.

The chromophore property of silver clusters has also been tested in living cancer cells by including such metal clusters in walls of polyelectrolyte-multilayer microcapsules [195]. A near-infrared laser pulse can remotely activate these capsules by thermal interactions with the silver clusters. This is demonstrated in Figure 6.17. The setup is illuminated by a 830 nm and 50 mW laser pulse for some seconds. One clearly observes the rupture of the capsule due to the strong coupling of the infrared laser with the silver clusters contained in its wall, through thermal processes. One can also note a sizable modification in the cancer cell, in particular at the side of its external membrane. Therefore this remote activation of such silver-capped capsules can be exploited for the release of encapsulated material inside living cells.

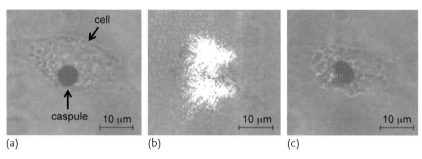

Figure 6.17 TEM images of a living cancer cell (gray structure in the background) containing a polyelectrolyte-multilayer capsule (dark circle) with silver clusters in its wall: before irradiation by a 830 nm and 50 mW laser (a), during irradiation (b) and after irradiation (c). Adapted from [195].

6.4.3.2 Cancer Therapies

The use of metal clusters in cancer therapies constitutes a growing and encouraging field in medicine. These new types of therapies offer lots of advantages as a less invasive treatment than conventional ones, accessibility to remote tumors which cannot be attained by surgery, and lead to fewer side effects on healthy cells than with standard radiotherapy, photothermal therapy or chemotherapy. The surface plasmon resonance of metal clusters is once again exploited here (see Section 2.2.3), in combination to the therapeutic window ranging from 700 to 1200 nm, by tuning the cluster's size and/or shape to position their plasmon peak in this frequency window. Therefore if one shines a near-infrared laser pulse on cancerous cells which are priorly loaded with designed gold clusters, one can very locally increase the temperature to a value sufficient to kill cancerous cells, whereas healthy ones can easily undergo such an elevation. For an exhaustive review of the heat generation and the thermophysical properties of gold clusters in biological media, see for example [196].

The use of gold nanostructures in photothermal therapies was first tested in 2003 on tumors in mice located at some millimeters below the skin [197]. More precisely, silica clusters of 110 nm of diameter are covered first by a 10 nm thick gold layer and second by polyethylene glycol polymers to passivate the designed nanostructures. By tuning the thickness of the gold nanoshell, the tailored nanoparticle exhibits a plasmon peak at 820 nm. The laser characteristics used in this experiment are then 820 nm, 4 W/cm^2, 5 mm spot diameter and exposure less than 6 min. The average increase of temperature can be as high as 35 K, well beyond the threshold for irreversible damage in living tissues. The irradiation without gold nanoshells raised the temperature only by about 10 K, which does not irreversibly damage cells. The depth profiles of the increase in temperature of irradiated mice tissues are presented in Figure 6.18. It is worth noting that laser-induced thermal therapies were first developed in the early 1990s, but without the use of a chromophore. The main limitation of these techniques at that time was their non-tissue selectivity, implying a large time of irradiation with not well-defined treated areas, and thus numerous side effects. The idea to couple the irradiation with a chromophore arose

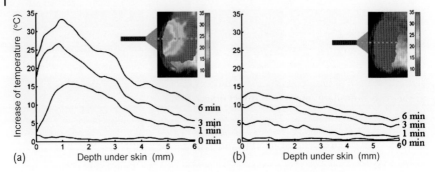

Figure 6.18 Increase of temperature as a function of depth from skin in mice after irradiation by a 820 nm and 4 W/cm² laser, after various irradiation time as indicated, with (a) and without gold-coated silica nanoparticles covered by polyethylene glycol (b). Adapted from [197].

in 1995 when a conventional fluorescent dye, the indocynanine green (ICG), was successfully tested [198]. Note however that the absorption in gold-silica nanoparticles is about 10^6 higher than in ICG, thus allowing the use of much less intense lasers and a shorter irradiation time.

The same idea of heating cancerous cells to destroy them is exploited in applying an external magnetic field on ferromagnetic clusters and is known as hyperthermia (for a review, see for example [199] and references therein). Indeed, one takes advantage of the energy loss proportional to the area delimited by the hysteresis curve. This loss in turn is converted into heat which can irreversibly damage tumor cells. The therapeutic threshold for destruction of a tumor is 42 °C sustained for 30 min. The very first application of such a technique was performed in 1957 using 20–100 nm Fe_2O_3 nanoparticles loaded in tissue samples and heated under a magnetic field of 1.2 MHz frequency. However the application to human patients is still under progress since some strong drawbacks still exist as muscle stimulation, in particular cardiac one and potentially arrhythmia, and appearance of Foucault currents in the body. Note also that in a physiological environment, oxidation of unprotected iron clusters causes a loss of their magnetic properties. There exist some attempts to protect them with gold layers. However the improvement of such a coating is not clear thus far [199]. Superparamagnetic clusters have also been explored, although they lack a hysteresis curve on the time scale of standard magnetic fields. However in high-frequency ac magnetic fields, such clusters do not have enough time to reorient their magnetization, and energy dissipation then becomes possible. Typical values of frequency and intensity of applied ac magnetic fields are 0.05–1.2 MHz and 0–15 kA/m.

The heating mechanisms involved in photothermal therapies or in hyperthermia are up to now described in terms of macroscopic variables governed by statistical physics and radiative transport equations, see for example [196]. From a microscopic point of view, the response of valence electrons in metal clusters to an external electromagnetic field is the predominant mechanism coming into play. At the side of the cell, double strand breaks of the nuclear DNA generally induce the death of the cell, since the latter one scarcely recovers from such a damage and consequently dies, while single-strand breaks are easily repaired. The application of an irradiation can directly ionize DNA and cause strand breaks. When a chromophore is used in addition, the energy loss from an excited cluster allows ionization of the surrounding molecules, giving rise to radicals or secondary low-energy electrons, which can in turn ionize DNA units. Indirect effects such as those just described are actually dominant (60%) over direct ones [200].

A very promising route for an efficient cancer chemoradiation relies on the concomitant use of the anticancer drug cisplatin (mentioned previously), 5 nm-diameter gold nanoparticles (GNP), and radiotherapy [201]. In this work, DNA films are either complexed with cisplatin alone, with GNP alone, or with both cisplatin and GNP, and then bombarded by 60 keV electrons. The results are presented in Figure 6.19. The number of DNA double strand breaks is measured without any chromophore. It is compared to the number of DNA double strand breaks in a mixture of GNP and DNA with a 1 : 1 molar ratio (black box), a mixture of cisplatin and DNA with a 2 : 1 molar ratio (gray box), and finally a mixture of the three of them (white box). An enhancement of the number of breaks is observed in each case. The case of cisplatin mixed alone with DNA, thanks to the Pt atom, exhibits a chromophore property since the number of double strand breaks increases by a factor of 2.5 when compared to the irradiation of DNA alone. The same effect has been observed when GNP are used instead of cisplatin. Interestingly, a 2 : 1 : 1 molar ratio between cisplatin, GNP and DNA gives rise to double strand breaks which are enhanced by a factor of 7.5 compared with the case of irradiation of pure DNA. The explanation of such an increase is more complex than a mere summation of the chromophore effect of cisplatin and that of GNP, and two-event processes. This enhancement demonstrates that a synergy between GNP and cisplatin exists at the side of the formation of double strand breaks, and corresponds to nontrivial electron transfers in the coupling of an irradiation and GNP-cisplatin-DNA complexes.

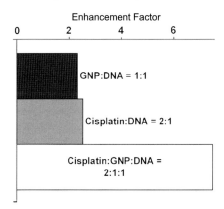

Figure 6.19 Double strand breaks from DNA films bombarded by 60 keV electrons. Plotted are the enhancement factors, compared with bare DNA, for a mixture of gold nanoparticles (GNP) and DNA (black box), cisplatin and DNA (gray box), and cisplatin, GNP and DNA (white box) with different molar ratios as indicated. Adapted from [201].

6.4.4
Nanotoxicity

In the previous sections, we have presented some of the numerous applications of clusters in medicine. Of course, in the case of medical imaging, the impact on cell viability has to be correctly studied to guarantee a safe use of such "devices" in a human body. The question of the cluster biological clearance is essential here because too high a cluster concentration can simply poison cells in a middle term and cause severe side effects. Clearance is more or less facilitated whether the cluster is taken by a cell or not. On the other hand, when clusters are used in a therapy, a cell uptake is desired, for instance to attach nuclear DNA and to kill the cell. One has then to balance the effects on cancerous cells with respect to the effects on healthy ones. Aside from the fact that clusters are more and more exploited, human beings are also exposed to nanoparticles produced in various industrial products or in manufacturing processes. Toxicity to living cells (cytotoxicity) is a question which must also be addressed but sometimes in a different context, for instance, when a person unintentionally inhales nanoparticles or uses a nanoparticle-containing sunscreen.

An exhaustive review on the cytotoxicity of carbon nanotubes, quantum dots and gold clusters can be found in [202]. The conclusions drawn there are however quite disturbing. Indeed, most of the studies have been carried in *in vitro* experiments, and it is not clear at all that the observed *in vitro* toxicities are relevant *in vivo*. Moreover, the clinical procedures are not aligned from one study to the next, rendering the comparison difficult, even impossible. Finally, the coating of a designed cluster can be as toxic as the cluster itself. What has been nevertheless observed is that the accumulation of clusters in a cell is one of the main factors of cytotoxicity.

Let us be a bit more specific and give some elements of cytotoxicity for each type of cluster previously discussed. At the side of fullerenes and carbon nanotubes, no clear and significant cytotoxicity of bare C_{60} has been reported [202]. However the combination of C_{60} with organic solvent as tetrahydrofuran (THF), used to enhance the preparation of aggregates of fullerenes, turns out to cause a sizable toxicity [203]. Following the same idea, single-wall carbon nanotubes (SWNT) assays do not clearly demonstrate a cytotoxicity at low concentration but the metal catalysts required to synthesize SWNT do strongly affect cell viability. Comparisons between SWNT and multiwall carbon nanotubes (MWNT) have also been carried out. From a cell point of view, MWNT exhibit a larger diameter (10–100 nm) than SWNT (some nm), and their possible internalization in a cell is certainly not the same [204]. The needle-like shape of MWNT also questions their impact in lung cells because of their superficial resemblance with asbestos. Cytotoxicity of commercial MWNT has been tested *in vivo* by injecting them in the abdominal cavity of mice [205]. An asbestos-like pathogenic behavior was reported. However this study did not address the risks of an inhalation exposure as such and inflammatory effects over the long term.

We finally turn to the case of metallic nanosystems, and first to gold nanoparticles (GNP). In general, few cytotoxicity effects are reported in the various existing studies [202]. For instance, different coatings of GNP (citrate, biotin, glucose, and cetyltrimethylammonium bromide (CTAB)) have been tested *in vitro* [206]. Despite a quick uptake of the surface-modified GNP into cells, no significant toxic effects were found. However this study also revealed that precursors of the nanoparticles such as $AuCl_4$ at the same concentration decreased the cell viability by 90%. We have also reported the use of gold nanoshells with a silica core in photothermal cancer therapies [197]. *In vivo* tests carried out in mice reported the distribution of polyethylene glycol gold-coated silica clusters and their clearance after therapeutic injection in blood, liver, lung, spleen, muscle, kidney, bone, brain tissues, and subcutaneous tumors [207]. It has been observed that the nanoparticles are quickly cleared from blood following an exponential law, while scavenged by organs. Interestingly, the maximum accumulation is obtained 24 h after injection in the kidney and tumor, whereas those in the other organs are one or two orders of magnitude lower. This indicates that 24 h after injection is the optimal time for photothermal subcutaneous tumor ablation. We finally indicate that the Au_3Cu_1 nanocapsules previously mentioned for magnetic resonant imaging (MRI) do not show any significant signal 4 h after injection and seem to be efficiently excreted in urine [185]. In addition, the lowest nanocapsule concentration used there retains a 100% mice viability 30 days after injection.

7
Conclusion

Our exploration of cluster science attempted to cover as many of the numerous facets as possible of a still emerging field of science. The domain is huge and we have certainly touched upon some aspects in a superficial manner, as well as overlooked many relevant issues. It was however our primary goal to give an overview and to motivate the reader to further pursue this rich field of science. We hope the manuscript managed to reach that goal. Looking back, we now try to draw a few conclusions and identify the general direction along which our discussions have developed.

One of the major issues is that cluster science is still in a young and expanding stage. Over the past few decades, it has grown into its own field of research, in which clusters themselves are recognized as objects of study. Many experimental and theoretical investigations were thus led in order to access the structure, and to a lesser (but growing) extent the dynamics of clusters. By clusters, we mean both isolated (free) clusters as well as clusters in contact with an environment, either deposited on a surface or embedded into a matrix or a solvent. Cluster science was strongly motivated by having clusters as unique objects which interpolate between atoms/molecules and bulk material. It is also remarkable how tight the connections are between cluster science and other fields of physics and chemistry. We have thus tried to illustrate this aspect in several places throughout this book, devoting a sizable portion of our discussions to interfaces with other fields of science. As examples of unexpected connections, we mention here biology and medicine, as well as, in a totally different context, astrophysics. But such "exotic" interfaces should not mask the more "conventional" and highly fruitful ones such as the connection with material science, a domain in which clusters do play a growing role. Cluster science is thus a field of its own and at the same time highly interdisciplinary, with many connections to other areas of study.

Another clear and important aspect of cluster science is the fact that it gathers very fundamental studies as well as highly applied ones. On the side of fundamental research, let us recall that clusters are, to date, amongst the most versatile objects to understand how matter builds up. The possibility of varying at will the size is crucial here. This provides, for example, a fascinating and challenging playground for the development of the theory of the many-body problem. This first holds from a structural point of view. However, recent experimental developments

An Introduction to Cluster Science, First Edition. Phuong Mai Dinh, Paul-Gerhard Reinhard, and Eric Suraud.
© 2014 WILEY-VCH Verlag GmbH & Co. KGaA. Published 2014 by WILEY-VCH Verlag GmbH & Co. KGaA.

show that details of cluster dynamics will soon be readily accessible, thus addressing key questions in our understanding of the dynamics of many-body systems as well. That even more demanding field certainly presents a challenge for the development of theory. It should be noted that theory development goes beyond mere cluster science which, in turn, is fruitful for both sides. The other facet of cluster science, namely its many applications, is also an essential feature of the field. It probably finds its most salient examples in intuitively remote areas such as biology and medicine. It is worth noting that these applications rely on very basic physical properties of the system, in particular its coupling to light.

Finally, and somewhat with the idea of addressing future challenges, we would like to emphasize that cluster science has reached a level at which one is progressively shifting from a mere observation to tailoring and even manipulation. This by no means implies that observation is at present obsolete. It is clear that we still need to pursue the many investigations on cluster properties themselves. But, as pointed out in many applications, we are already at a stage where we know enough about the nature of clusters to be able to tailor and even manipulate them. We have seen several examples of applications in biology and medicine, for example in relation to biocompatibility. We have also seen on this occasion that tuning cluster size, for example, opens the door to the manipulation of clusters and/or composites of clusters and biologically active molecules. The same holds true in materials science, for example, with the case of clusters deposited on a surface. These various applications are mostly handled from a yet practical point of view. There thus remains for sure a number of investigations to be performed on such applications in order to better understand the underlying physical and chemical mechanisms at play.

Further Reading

We list here a few textbooks used in the preparation of this manuscript, as well as the proceedings of a few major conferences on cluster physics. The latter list is chronological and by no means exhaustive.

General physics

- Jackson, J. D., (1962) *Classical Electrodynamics*, John Wiley & Sons, New York
- Messiah, A., (1970) *Quantum Mechanics Vol. I*, John Wiley & Sons, New York

General many-body theory

- Pines, D. and Nozières, P. (1966) *The theory of Quantum Liquids*, Benjamin, New York.
- Fetter, A. L. and Walecka, J. D. (1971) *Quantum theory of Many-Particle Systems*, McGraw-Hill, New York.
- Parr, R. G. and Yang, W. (1989) *Density-Functional Theory of Atoms and Molecules*, Oxford University Press, Oxford.
- Dreizler, R. M. and Gross, E. K. U. (1990) *Density Functional Theory: An Approach to the Quantum Many-Body Problem*, Springer, Berlin.
- Mahan, G. D. (1993) *Many Particle Physics*, Plenum Press, New York.
- Krotscheck, E. and Navarro, J. (2002) *Microscopic Approaches to Quantum Liquids in Confined Geometries*, World Scientific, Singapore.
- Lipparini, E. (2008) *Modern Many-Particle Physics (2nd edn)*, World Scientific, Singapore.
- Maruhn, J., Reinhard, P.-G., and Suraud, E. (2010) *Simple Models of Many-Fermion Systems*, Springer, Heidelberg.

Solid state physics

- Ashcroft, N. W. and Mermin, N. D. (1976) *Solid State Physics*, Saunders College, Philadelphia.
- Callaway, J. (1991) *Quantum Theory of the Solid State*, Academic, London.

- Elliott, S. (1998) *The Physics and Chemistry of Solids*, John Wiley & Sons, New York.

Atomic and molecular physics – chemistry

- Brandsen, R. H. (1983) *Atomic Collision Theory*, Benjamin, Reading.
- Szasz, L. (1985) *Pseudopotential Theory of Atoms and Molecules*, John Wiley & Sons, New York.
- Faisal, F. H. M. (1987) *Theory of Multiphoton Processes*, Plenum Press, New York.
- Yamaguchi, Y., Osamure, Y., Goddard, J. D., and Schaefer III, H. F. (1994) *A New Dimension to Quantum Chemistry. Analytic Derivative Methods in Ab-Initio Electronic Structure Theory*, Oxford University Press, Oxford.
- Zewail, A. H. (1994) *Femtochemistry, Vol. I and II*, World Scientific, Singapore.
- Pettifor, D., (1995) *Bonding and Structure of Molecules and Solids*, Clarendon, Oxford.
- Bransden, B. H. and Joachain, J., (1997) *Physics of Atoms and Molecules*, Longman, London.
- Pauly, H., (2000) *Atom, Molecule, and Cluster Beams I and II*, Springer, Berlin.

Cluster physics

- Haberland, H., (1994) *Clusters of Atoms and Molecules: 1 – Theory, Experiment, and Clusters of Atoms;2 – Solvation and Chemistry of Free Clusters, and Embedded, Supported and Compressed Clusters*, Vol. 52 and 56, Springer Series in Chemical Physics, Berlin.
- Sugano, S. and Koizumi, H. (1998) *Microcluster Physics (2nd edn)*, Springer, Berlin.
- Ekardt, W. (1999) *Metal Clusters*, John Wiley & Sons, New York.
- Jellinek, J. (1999) *Theory of Atomic and Molecular Clusters*, Springer, Berlin.
- Reinhard, P.-G. and Suraud, E. (2003) *Introduction to Cluster Dynamics*, Wiley-VCH Verlag GmbH, Berlin.
- Alonso, J. A. (2005) *Structure and Properties of Atomic Nanoclusters*, Imperial College Press, London.
- Castleman Jr., A. W., Berry, R. S., Haberland, H., Jortner, J., and Kondow, T., (eds) (since 1999) *Springer Series in Cluster Physics*, Springer, Heidelberg. http://www.springer.com/series/4080 (accessed 11 August 2013).
- *Proceedings of the ISSPIC conferences of the past decade:*
  ISSPIC 11: Eur. Phys. J. D **24** (2003); ISSPIC 12: Eur. Phys. J. D **34** (2005); ISSPIC 13: Eur. Phys. J. D **43** (2007); ISSPIC 14: Eur. Phys. J. D **52** (2009); ISSPIC 15: Eur. Phys. J. D **63** (2011); ISSPIC 16: Eur. Phys. J. D **67** (2013)

References

1. Suessmann, G. (1963) *Einführung in die Quantenmechanik I*. Bibliographisches Institut, Mannheim.
2. Lehmann, J., Merschdorf, M., Pfeiffer, W., Thon, A., Voll, S., and Gerber, G. (2000) *Phys. Rev. Lett.*, **85**, 2921.
3. Mie, G. (1908) *Ann. Phys. (Leipz.)*, **25**, 377.
4. Kroto, H.W. (1987) *Nature*, **329**, 529.
5. Knight, W.D., Clemenger, K., de Heer, W.A., Saunders, W.A., Chou, M.Y., and Cohen, M.L. (1984) *Phys. Rev. Lett.*, **52**, 2141.
6. Kaiser, A.B. (2001) *Rep. Prog. Phys.*, **64**, 1.
7. Krausz, F. and Ivanov, M. (2009) *Rev. Mod. Phys.*, **81**, 163.
8. Lifschitz, E.M. and Pitajewski, L.P. (1981) *Physical Kinetics: Course of Theoretical Physics*. Butterworth-Heinemann, Oxford.
9. Bransden, B.H. and Joachain, C.J. (2003) *Physics of Atoms and Molecules*. Prentice Hall, Upper Saddle River.
10. Ashcroft, N.W. and Mermin, N.D. (1976) *Solid State Physics*. Saunders College, Philadelphia.
11. Cheshnovsky, O., Taylor, K.J., Conceicao, J., and Smalley, R.E. (1990) *Phys. Rev. Lett.*, **64**, 1785.
12. Kreibig, U. and Vollmer, M. (1993) *Optical Properties of Metal Clusters*, Springer Series in Materials Science. Springer, Berlin.
13. Brechignac, C., Busch, H., Cahuzac, P., and Leygnier, J. (1994) *J. Chem Phys.*, **101**, 6992.
14. Haberland, H. (ed.) (1999) *Clusters of Atoms and Molecules 1 – Theory, Experiment, and Clusters of Atoms*. Springer Series in Chem. Physics. Springer, Berlin.
15. Meiwes-Broer, K.-H. (ed.) (2000) *Metal Clusters at Surfaces*. Springer, Berlin.
16. Binns, C. (2001) *Surf. Sci. Rep.*, **44**, 1.
17. de Heer, W.A. (1993) *Rev. Mod. Phys.*, **65**, 611.
18. Milani, P. and Iannotta, S. (1999) *Cluster Beam Synthesis of Nanostructured Materials*. Springer, Berlin.
19. Pauly, H. (2000) *Atom, Molecule, and Cluster Beams I and II*. Springer, Berlin.
20. Larson, R.A., Ncoh, S.K., and Herschbach, D.R. (1984) *Rev. Sci. Instrum.*, **45**, 1511.
21. Näher, U., Björnholm, S., Frauendorf, S., Garcias, F., and Guet, C. (1997) *Phys. Rep.*, **285**, 245.
22. Ellert, C., Schmidt, M., Schmitt, C., Reiners, T., and Haberland, H. (1995) *Phys. Rev. Lett.*, **75**, 1731.
23. Brandsen, R.H. (1983) *Atomic Collision Theory*. Benjamin, Reading.
24. Bransden, B.H. and McDowell, M.R.C. (1992) *Charge Exchange and Theory of Ion-Atom Collisions*. Clarendon, Oxford.
25. Goldberger, M.L. and Watson, K.M. (1964) *Collision Theory*. John Wiley & Sons, New York.
26. Zewail, A.H. (1994) *Femtochemistry, Vol. I and II*. World Scientific, Singapore.
27. Gasiorowicz, S. (1974) *Quantum Physics*. John Wiley & Sons, New York.
28. Maruhn, J.A., Reinhard, P.-G. and Suraud, E. (2010) *Simple Models of Many-Fermion Systems*. Springer, Heidelberg.
29. Irvine, J.M. (1972) *Nuclear Structure Theory*. Clarendon Press, Oxford.

30. Brack, M. (1993) *Rev. Mod. Phys.*, **65**, 677.
31. Rohlfing, E.A., Cox, D.M., and Kaldor, A. (1984) *J. Chem. Phys.*, **81**, 3322.
32. Haken, H. and Wolf, H.C. (2000) *The Physics of Atoms and Quanta*. Springer, Berlin.
33. Billas, I.M.L., Châtelain, A., and de Heer, W.A. (1994) *Science*, **265**, 1682.
34. Haberland, H. and Schmidt, M. (1999) *Eur. Phys. J. D*, **6**, 109.
35. Nesterenko, V.O., Kleinig, W., and Reinhard, P.-G. (2002) *Eur. Phys. J. D*, **19**, 57.
36. Lamour, E., Prigent, C., Rozet, J.P., and Vernhet, D. (2007) *J. Phys. Conf. Ser.*, **88**, 012035.
37. Kostko, O., Bartels, C., Schwobel, J., Hock, C., and von Issendorff, B. (2007) *J. Phys. : Conf. Ser.*, **88**, 012034.
38. Bordas, C., Paulig, F., Heln, H., and Huestis, D.L. (1996) *Rev. Sci. Instrum.*, **67**, 2257.
39. Bartels, C., Hok, C., Huwer, J., Kuhnen, R., Schwöbel, J., and von Issendorff, B. (2009) *Science*, **323**, 132.
40. Born, M. and Huang, K. (1954) *Dynamical Theory of Crystal Lattices*. Oxford University Press, Oxford.
41. Weissbluth, M. (1978) *Atoms and Molecules*. Academic Press, San Diego.
42. Garraway, B.M. and Suominen, K.-A. (1995) *Rep. Prog. Phys.*, **58**, 365.
43. Grisenti, R.E., Schöllkopf, W., Toennies, J.P., Hegerfeldt, G.C., Köhler, T., and Stoll, M. (2000) *Phys. Rev. Lett.*, **85**, 2284.
44. Car, R. and Parrinello, M. (1985) *Phys. Rev. Lett.*, **55**, 2471.
45. Hutter, J. (2012) *Wiley Interdisciplinary Reviews: Computational Molecular Science*, **2**, 604.
46. Calvayrac, F., Reinhard, P.-G., Suraud, E., and Ullrich, C.A. (2000) *Phys. Rep.*, **337**, 493.
47. Reinhard, P.-G. and Suraud, E. (2003) *Introduction to Cluster Dynamics*. John Wiley & Sons, New York.
48. Tully, J.C. (1990) *J. Chem. Phys.*, **93**, 1061.
49. Ferguson, D.M., Siepmann, J.I., and Truhlar, D.G. (2009) *Monte Carlo Methods in Chemical Physics*. John Wiley & Sons.
50. Spiegelmann, F. and Poteau, R. (1995) *Comments At. Mol. Phys.*, **31**, 395.
51. Callaway, J. (1991) *Quantum Theory of the Solid State*. Academic, London.
52. Mahan, G.D. (1993) *Many Particle Physics*. Plenum Press, New York.
53. Vollhardt, D. (1984) *Rev. Mod. Phys.*, **56**, 99.
54. Garcia, M.E., Pastor, G.M., and Benneman, K.H. (1991) *Phys. Rev. Lett.*, **67**, 1142.
55. Proykova, A., Nikolova, D., and Berry, R.S. (2002) *Phys. Rev. B*, **65**, 085411.
56. Ercolessi, F.: A molecular dynamics primer, http://www.freescience.info/framepage.php?link=http://www.fisica.uniud.it/~ercolessi/md/ (accessed 11 August 2013).
57. Cherchneff, I. and Dwek, E. (2009) *Astrophys.*, **703**, 642.
58. Vlasov, A.A. (1950) *Many Particle Theory and Its Applications to Plasma*. Gordon and Breach, New York.
59. Kramer, P. and Saraceno, M. (1967) *Lecture Notes in Physics*, 140.
60. Madjet, M., Guet, C., and Johnson, W.R. (1995) *Phys. Rev. A*, **51**, 1327.
61. Lyalin, A., Solov'yov, A., and Greiner, W. (2002) *Phys. Rev. A*, **65**, 043202.
62. Yamaguchi, Y., Osamure, Y., Goddard, J.D., and Schaefer III, H.F. (1994) *A New Dimension to Quantum Chemistry*. Oxford University Press, Oxford.
63. Bonačic-Koutecký, V., Fantucci, P., and Koutecký, J. (1991) *Chem. Rev.*, **91**, 1035.
64. Bonačic-Koutecký, V., Pittner, J., Fuchs, C., Fantucci, P., Guest, M.F., and Koutecký, J. (1996) *J. Chem. Phys.*, **104**, 1427.
65. Jordan, G. and Scrinzi, A. (2009) Strongly driven few-fermion systems – MCTDHF, in *Multidimensional Quantum Dynamics: MCTDH Theory and Applications*, John Wiley & Sons, Berlin. Chapter 16, p. 161.
66. Ceperley, D.M. and Alder, B.J. (1980) *Phys. Rev. Lett.*, **45**, 566.
67. Needs, R.J., Kent, P.R.C., Porter, A.R., Towler, M.D., and Rajagopal, G. (2002) *Int. J. Quantum Chem.*, **86**, 218.
68. Parker, J., Taylor, K.T., Clark, C., and Blodgett-Ford, S. (1996) *J. Phys. B*, **29**, L33.

69 Nishioka, H., Hansen, K., and Mottelson, B.R. (1990) *Phys. Rev. B*, **42**, 9377.
70 Hohenberg, P. and Kohn, W. (1964) *Phys. Rev.*, **136**, 864.
71 Kohn, W. and Sham, L.J. (1965) *Phys. Rev.*, **140**, 1133.
72 Dreizler, R.M. and Gross, E.K.U. (1990) *Density Functional Theory: An Approach to the Quantum Many-Body Problem*. Springer, Berlin.
73 Gross, E.K.U., Dobson, J.F., and Petersilka, M. (1996) *Top. Curr. Chem.*, **181**, 81.
74 Parr, R.G. and Yang, W. (1989) *Density-Functional Theory of Atoms and Molecules*. Oxford University Press, Oxford.
75 Ashcroft, N.W. and Langreth, D.C. (1967) *Phys. Rev.*, **155**, 682.
76 Perdew, J.P. and Wang, Y. (1992) *Phys. Rev. B*, **45**, 13244.
77 Ekardt, W. (1984) *Phys. Rev. Lett.*, **52**, 1925.
78 Kümmel, S. and Kronik, L. (2008) *Rev. Mod. Phys.*, **80**, 3.
79 Zwicknagel, G., Toepffer, C., and Reinhard, P.-G. (1999) *Phys. Rep.*, **309**, 117.
80 Ditmire, T., Tisch, J.W.G., Springate, E., Mason, M.B., Hay, N., Marangos, J.P., and Hutchinson, M.H.R. (1997) *Phys. Rev. Lett.*, **78**, 2732.
81 Bertsch, G.F. and Das Gupta, S. (1988) *Phys. Rep.*, **160**, 190.
82 Bonasera, A., Gulminelli, F., and Molitoris, J. (1994) *Phys. Rep.*, **243**, 1.
83 Abe, Y., Ayik, S., Reinhard, P.-G., and Suraud, E. (1996) *Phys. Rep.*, **275**, 49.
84 Fennel, T., Meiwes-Broer, K.-H., Tiggesbäumker, J., Dinh, P.M., Reinhard, P.-G., and Suraud, E. (2010) *Rev. Mod. Phys.*, **82**, 1793.
85 Szasz, L. (1985) *Pseudopotential Theory of Atoms and Molecules*. John Wiley & Sons, New York.
86 Abarenkov, I.V. and Heine, V. (1965) *Philos. Mag.*, **12**, 529.
87 Aziz, R.A. and Chen, H.H. (1977) *J. Chem. Phys.*, **67**, 5719.
88 Reinhard, P.-G., Weisgerber, S., Genzken, O., and Brack, M. (1994) *Z. Phys. A*, **349**, 219.
89 Alasia, F., Serra, L., Broglia, R.A., Giai, N.V., Lipparini, E., and Roman, H.E. (1995) *Phys. Rev. B*, **52**, 8488.
90 Nitzan, A. (2001) *Anu. Rev. Phys. Chem.*, **52**, 681.
91 Press, W.H., Teukolsky, S.A., Vetterling, W.T., and Flannery, B.P. (1992) *Numerical Recipes*. Cambridge University Press, Cambridge.
92 Weisskopf, V. (1937) *Phys. Rev.*, **52**, 295.
93 Balian, R. and Bloch, C. (1974) *Ann. Phys. (NY)*, **85**, 514.
94 Brack, M. and Bhaduri, R.K. (1997) *Semiclassical Physics*. Addison-Wesley, Reading.
95 Martin, T.P. (1993) *Phys. Rep.*, **273**, 199.
96 Englman, R. (1972) *The Jahn-Teller Effect in Molecules and Crystals*. John Wiley & Sons, London.
97 Reiners, T., Ellert, C., Schmidt, M., and Haberland, H. (1995) *Phys. Rev. Lett.*, **74**, 1558.
98 Babst, J. and Reinhard, P.-G. (1997) *Z. Phys. D*, **42**, 209.
99 Borggreen, J., Chowdhury, P., Kebaili, N., Lundsberg-Nielsen, L., Luetzenkirchen, K., Nielsen, M.B., Pedersen, J., and Rasmussen, H.D. (1993) *Phys. Rev. B*, **48**, 17507.
100 Ellert, C., Schmidt, M., Haberland, H., Veyret, V., and Bonačic-Koutecký, V. (2002) *J. Chem. Phys.*, **117**, 3711.
101 Yannouleas, C. and Broglia, R. (1992) *Ann. Phys. (NY)*, **217**, 105.
102 Schlipper, R., Kusche, R., von Issendorff, B., and Haberland, H. (2001) *Appl. Phys. A*, **72**, 255.
103 Faisal, F.H.M. (1987) *Theory of Multiphoton Processes*. Plenum Press, New York.
104 Wopperer, P., Faber, B., Dinh, P.M., Suraud, E., and Reinhard, P.-G. (2012) *Phys. Rev. A*, **85**, 015402.
105 Brabec, T. and Krausz, F. (2000) *Rev. Mod. Phys.*, **72**, 545.
106 Zuo, T. and Bandrauk, A.D. (1995) *Phys. Rev. A*, **52**, R2511.
107 Ditmire, T., Zweiback, J., Yanovsky, V.P., Cowan, T.E., Hays, G., and Wharton, K.B. (1999) *Nature*, **398**, 489.
108 Li, S.H., Wang, C., Zhu, P.P., Wang, X.X., Li, R.X., Ni, G.Q., and Xu, Z.Z. (2003) *Chin. Phys. Lett.*, **20**, 1247.
109 Lebeault, M.A., Viallon, J., Chevaleyre, J., Ellert, C., Normand, D., Schmidt, M., Sublemontier, O., Guet, C., and Huber, B. (2002) *Eur. Phys. J. D*, **20**, 233.

110 Teuber, S., Döppner, T., Fennel, T., Tiggesbäumker, J., and Meiwes-Broer, K.H. (2001) *Eur. Phys. J. D*, **16**, 59.

111 Fukuda, Y., Yamakawa, K., Akahane, Y., Aoyama, M., Inoue, N., Ueda, H. and Kishimoto, Y. (2003) *Phys. Rev. A*, **67**, 061201R.

112 Meiwes-Broer, K.-H. and Berndt, R. (eds) (2007) *Atomic Clusters at Surfaces and in Thin Films*, vol. 45. Eur. Phys. J. D, topical issue, Springer, Heidelberg.

113 Ouacha, H., Hendrich, C., Hubenthal, F., and Träger, F. (2005) *Appl. Phys. B*, **81**, 663.

114 Sanchez, A., Abbet, S., Heiz, U., Schneider, W.-D., Häkkinen, H., Barnett, R.N., and Landman, U. (1999) *J. Phys. Chem. A*, **103**, 9573.

115 Janes, D.B., Batistuta, M., Datta, S., Melloch, M.R., Andres, R.P., Liu, J., Chen, N.-P., Lee, T., Reifenberger, R., Chen, E.H., and Woodall, J.M. (2000) *Superlattices and Microstructures*, **27**, 555.

116 Schaadt, D.M., Feng, B., and Yu, E.T. (2005) *Appl. Phys. Lett.*, **86**, 063106.

117 Richardson, H.H., Hickman, Z.N., Govorov, A.O., Thomas, A.C., Zhang, W., and Kordesch, M.E. (2006) *Nano Lett.*, **6**, 783.

118 Niv, M.Y., Bargheer, M., and Gerber, R.B. (2000) *J. Chem. Phys.*, **113**, 6660.

119 Warshel, A. and Levitt, M. (1976) *J. Mol. Biol.*, **103**, 227.

120 Rösch, N., Nasluzov, V.A., Neyman, K.M., Pacchioni, G., and Vayssilov, G.N. (2004) Supported Metal Species and Adsorption Complexes on Metal Oxides and in Zeolites: Density Functional Cluster Model Studies, in: *Computational Material Science*, Theoretical and Computational Chemistry Series, vol. 15 (ed. J. Leszczynski), Elsevier, Amsterdam, pp. 367–450.

121 Diederich, T., Döppner, T., Tiggesbäumker, J., and Meiwes-Broer, K.-H. (2001) *Phys. Rev. Lett.*, **86**, 4807.

122 Krischok, S., Stracke, P., and Kempter, V. (2006) *Appl. Phys. A*, **82**, 167.

123 Seifert, G., Kaempfe, M., Berg, K.-J., and Graener, H. (2000) *Appl. Phys. B*, **71**, 795.

124 Liu, B., Nielsen, S.B., Hvelplund, P., Zettergren, H., Cederquist, H., Manil, B., and Huber, B.A. (2006) *Phys. Rev. Lett.*, **97**, 133401.

125 Enders, A., Skomski, R., and Honolka, J. (2010) *J. Phys. Condensed Mater.*, **22**, 433001.

126 Stienkemeier, F. and Vilesov, A.F. (2001) *J. Chem. Phys.*, **115**, 10119.

127 Döppner, T., Fennel, T., Radcliffe, P., Tiggesbäumker, J., and Meiwes-Broer, K.-H. (2006) *Phys. Rev. A.*, **73**, 031202R.

128 Döppner, T., Fennel, T., Diederich, T., Tiggesbäumker, J., and Meiwes-Broer, K.-H. (2005) *Phys. Rev. Lett.*, **94**, 013401.

129 Bruhwiler, D., Leiggener, C., Glaus, S. and Calzaferri, G. (2002) *J. Phys. Chem.*, **106**, 3770, 2002.

130 Halas, N.J., Lal, S., Chang, W.-S., Link, S., and Nordlander, P. (2011) *Chem. Rev.*, **111**, 3913.

131 Pavlyukh, Y., Berakdar, J., and Hübner, W. (2010) *Phys. Status Solidi (b)*, **247**, 1056.

132 Gresh, N., Parisel, O., and Giessner-Prettre, C. (1999) *THEOCHEM*, **458**, 27.

133 Vreven, T. and Morokuma, K. (2000) *J. Comput. Chem.*, **21**, 1419.

134 Torrent, M., Vreven, T., Musaev, D.G., Morokuma, K., Farkas, O., and Schlegel, H.B. (2002) *J. Am. Chem. Soc.*, **124**, 192.

135 Dinh, P.M., Reinhard, P.-G., and Suraud, E. (2010) *Phys. Rep.*, **485**, 43.

136 Park, H., Park, J., Lim, A.K.L., Anderson, E.H., Alivisatos, A.P., and McEuen, P.L. (2000) *Nature*, **407**, 57.

137 Cleland, A.N. (ed.) (2003) *Foundations of Nanomechanics*. Springer, Berlin.

138 Jia, C.-J. and Schüth, F. (2011) *Phys. Chem.*, **13**, 2457.

139 Louis, C., and Pluchery, O. (2012) *Gold nano particles for physics, chemistry and biology*. World Scientific, Singapore.

140 Harding, C., Habibpour, V., Kunz, S., Farnbacher, A.N.-S., Yoon, U.H.B., and Landman, U. (2009) *J. Am. Chem. Soc.*, **131**, 538.

141 Primo, A., Corma, A., and Garcia, H. (2011) *Phys. Chem. Chem. Phys.*, **13**, 886.

142 Centeno, M.A., Hidalgo, M.C., Dominguez, M.I., Navio, J.A., and Odriozola, J.A. (2003) *Catal. Lett.*, **123**, 198.

143 Xu, R., Wang, D., Zhang, J., and Li, Y. (2006) *Chem. Asian J.*, **1**, 888.

144 Cuenya, B.R. (2010) *Thin Solid Films*, **518**, 3127.
145 Reinhard, P.-G. and Suraud, E. (1998) *Eur. Phys. J. D*, **3**, 175.
146 Lee, K.-G., Eghlidi, H., Chen, X.-W., Renn, A., Götzinger, S., and Sandoghdar, V. (2012) *Opt. Express*, **20**, 23331.
147 Eklund, P.C. and Rao, A.M. (eds) (1999) *Fullerene Polymers and Fullerene Polymer Composites*. Springer Series in Materials Science. Springer, Heidelberg.
148 Claridge, S.A., Jr, A.W.C., Khanna, S.N., Murray, C.B., Sen, A., and Weiss, P.S. (2009) *ACS Nano*, **3**, 244.
149 Reber, A.C., Khanna, S.N., and Jr, A.W.C. (2007) *J. Am. Chem. Soc.*, **129**, 10189.
150 Bohr, Å. and Mottelson, B.R. (1980) *Struktur der Atomkerne II. Kerndeformationen*. Akademie Verlag, Berlin.
151 Greiner, W. and Maruhn, J.A. (2008) *Nuclear models*. Springer, Berlin Heidelberg.
152 Glendenning, N.K. (1997) *Compact Stars*. Springer, Heidelberg.
153 Raman, S., Jr, A.W.C., Kahane, S., and Bhatt, K.H. (1989) *At. Data Nucl. Data Table*, **42**, 1.
154 Brack, M., Damgård, J., Jensen, A.S., Pauli, H.C., Strutinsky, V.M., and Wong, C.Y. (1972) *Rev. Mod. Phys.*, **44**, 320.
155 Reimann, S.M., and Manninen, M. (2002) *Rev. Mod. Phys.*, **74**, 1283.
156 Eisenberg, J.H. and Greiner, W. (1970) *Excitation Mechanims of the Nuclei*. Volume 2 of *Nuclear Theory*. North Holland, Amsterdam.
157 van der Woude, A. (1991) The electric giant rersonances, in *Electric and Magnetic Giant Resonances in Nuclei*, vol. 7 (ed. J. Speth). Int. Rev. Nucl. Phys. World Scientific, pp. 99.
158 Bohigas, O., Lane, A.M., and Martorell, J. (1979) *Phys. Rep.*, **51**, 267.
159 Cherchneff, I. and Dwek, E. (2010) *Astrophys.*, **713**, 1.
160 Russell, R.W., Soifer, B.T., and Willner, S.P. (1977) *Astrophys.*, **217**, L149.
161 Tielens, A.G.G.M. (2008) *Anu. Rev. Astron. Astrophys.*, **46**, 289.
162 Draine, B.T. and Li, A. (2007) *Astrophys.*, **657**, 810.
163 Kwok, S. and Zhang, Y. (2011) *Nature*, **479**, 80.
164 Duley, W.W., Grishko, V.I., and Pinho, G. (2006) *Astrophys.*, **642**, 966.
165 Li, A. and Draine, B.T. (2012) *Astrophys. Lett.*, **760**, L35.
166 Twomey, S. (1974) *Atmos. Environ.*, **8**, 1251.
167 Kabatas, B., Menzel, W.P., Bilgili, A., and Gumley, L.E. (2013) *J. Appl. Meteor. Climatol.*, **52**, 230.
168 Aitken, J.A. (1897) *Trans. R. Soc.*, **39**, 15.
169 Asmi, A., Wiedensohler, A., Laj, P., Fjaeraa, A.-M., Sellegri, K., Birmili, W., Weingartner, E., Baltensperger, U., Zdimal, V., Zikova, N., Putaud, J.-P., Marinoni, A., Tunved, P., Hansson, H.-C., Fiebig, M., Kivekäs, N., Lihavainen, H., Asmi, E., Ulevicius, V., Aalto, P.P., Swietlicki, E., Kristensson, A., Mihalopoulos, N., Kalivitis, N., Kalapov, I., Kiss, G., de Leeuw, G., Henzing, B., Harrison, R.M., Beddows, D., O'Dowd, C., Jennings, S.G., Flentje, H., Weinhold, K., Meinhardt, F., Ries, L., and Kulmala, M. (2011) *Atmos. Chem. Phys.*, **11**, 5505.
170 Kulmala, M., Vehkamäki, H., Petäjä, T., Maso, M.D., Lauri, A., Kerminen, V.-M., Birmili, W., and McMurry, P.H. (2004) *J. Aerosol Sci.*, **35**, 143.
171 Kulmala, M. and Kerminen, V.-M. (2008) *Atmos. Res.*, **90**, 132.
172 Pierce, J.R. and Adams, P.J. (2007) *Atmos. Chem. Phys.*, **7**, 1367.
173 Weber, R.J., Marti, J.J., McMurry, P.H., Eisele, F.L., Tanner, D.J., and Jefferson, A. (1997) *J. Geophys. Res.*, **102**, 4375.
174 Hvelplund, P., Kurtén, T., Støchkel, K., Ryding, M.J., Nielsen, S.B., and Uggerud, E. (2010) *J. Phys. Chem. A*, **114**, 7301.
175 Stewart, J.J.P. (2007) *Molecular Orbital Package (MOPAC)*, http://openmopac.net (accessed 11 August 2013).
176 Salby, M.L. (2012) *Physics of the Atmosphere and Climate*. Cambridge University Press, Cambridge.
177 Koehler, H. (1936) *Trans. Faraday Soc.*, **32**, 1152.
178 Petters, M.D. and Kreidenweis, S.M. (2007) *Atmos. Chem. Phys.*, **7**, 1961.
179 Riehemann, K., Schneider, S., Luger, T., Godin, B., Ferrari, M., and Fuchs, H. (2009) *Angew. Chem. Int. Ed.*, **48**, 872.

180 Stark, W.J. (2011) *Angew. Chem. Int. Ed.*, **50**, 1242.
181 Dykman, L. and Khlebtsov, N. (2012) *Chem. Soc. Rev.*, **41**, 2256.
182 Liz-Marzán, L.M., Giersig, M., and Mulvaney, P. (1996) *Langmuir*, **12**, 4329.
183 Decuzzi, P. and Ferrari, M. (2008) *Biomat.*, **29**, 377.
184 Liu, C.H., Huang, S., Cui, J., Kim, Y.R., Farrar, C.T., Moskowitz, M.A., Rosen, B.R., and Liu, P.K. (2007) *FASEB Journal*, **21**, 3004.
185 Su, C.-H., Sheu, H.-S., Lin, C.-Y., Huang, C.-C., Lo, Y.-W., Pu, Y.-C., Weng, J.-C., Shieh, D.-B., Chen, J.-H., and Yeh, C.-S. (2007) *J. Am. Chem. Soc.*, **129**, 2139.
186 Liu, G.L., Long, Y.-T., Choi, Y., Kang, T. and Lee, L.P. (2007) *Nat. Methods*, **4**, 1015.
187 Choi, Y., Kang, T., and Lee, L.P. (2009) *Nano Lett.*, **9**, 85.
188 Balasubramaniana1, K., Kurkinaa1, T., Ahmada, A., Burgharda, M., and Kern, K. (2012) *J. Mater. Res.*, **27**, 391.
189 Kurkina, T., Vlandas, A., Ahmad, A., Kern, K., and Balasubramanian, K. (2011) *Angew. Chem. Int. Ed.*, **50**, 3710.
190 Doane, T.L. and Burda, C. (2012) *Chem. Soc. Rev.*, **41**, 2885.
191 Cheng, Y., Samia, A. C, Meyers, J.D., Panagopoulos, I., Fei, B., and Burda, C. (2008) *J. Am. Chem. Soc.*, **130**, 10643.
192 Dhar, S., Daniel, W.L., Giljohann, D.A., Mirkin, C.A., and Lippard, S.J. (2009) *J. Am. Chem. Soc.*, **131**, 14652.
193 Sershen, S.R., Westcott, S.L., Halas, N.J., and West, J.L. (2000) *J. Biomed. Mater. Res.*, **51**, 293.
194 Roggan, A., Friebel, M., Dörschel, K., Hahn, A., and Müller, G. (1999) *J. Biomed. Opt.*, **4**, 36.
195 Skirtach, A.G., Muñoz Javier, A., Kreft, O., Köhler, K., Piera Alberola, A., Möhwald, H., Parak, W.J., and Sukhorukov, G.B. (2006) *Angew. Chem. Int. Ed.*, **45**, 4612.
196 Qin, Z. and Bischof, J.C. (2012) *Chem. Soc. Rev.*, **41**, 1191.
197 Hirsch, L.R., Stafford, R.J., Bankson, J.A., Sershen, S.R. Rivera, B., Price, R.E., Hazle, J.D., Halas, N.J., and West, J.L. (2003) *Proc. Natl. Acad. Sci. (USA)*, **100**, 13549.
198 Chen, W.R., Adams, R.L., Bartels, K.E., and Nordquist, R.E. (1995) *Cancer Lett.*, **94**, 125, 1995.
199 Huber, D.L. (2005) *Small*, **1**, 482.
200 Sanche, L. (2010) Low-energy electron interaction with DNA: Bond dissociation and formation of transient anions, radicals, and radical anions., in *Radical and Radical Ion Reactivity in Nucleic Acid Chemistry*, John Wiley & Sons, Chapter 9, p. 239.
201 Zheng, Y. and Sanche, L. (2009) *Radiat. Res.*, **172**, 114.
202 Lewinski, N., Colvin, V., and Drezek, R. (2008) *Small*, **4**, 26.
203 Kovochich, M., Espinasse, B., Auffan, M., Hotze, E.M., Wessel, L., Xia, T., Nel, A.E., and Wiesner, M.R. (2009) *Environ. Sci. Technol.*, **43**, 6378.
204 Jia, G., Wang, H., Yan, L., Wang, X., Pei, R., Yan, T., Zhao, Y., and Guo, X. (2005) *Environ. Sci. Technol.*, **39**, 1378.
205 Poland, C.A., Duffin, R., Kinloch, I., Maynard, A., Wallace, W.A.H., Seaton, A., Stone, V., Brown, S., MacNee, W., and Donaldson, K. (2008) *Nat. Nanotechnol.*, **3**, 423.
206 Connor, E.E., Mwamuka, J., Gole, A., Murphy, C.J., and Wyatt, M.D. (2005) *Small*, **1**, 325.
207 James, W.D., Hirsch, L.R., West, J.L., O'Neal, P.D., and Payne, J.D. (2007) *J. Radioanal. Nucl. Chem.*, **271**, 455.

Index

Symbols
β_2, 91, 92
C_{60}, 16, 25, 37, 93, 120, 156
κ-Köhler theory, 143

A
ab initio, 55
abundance spectrum, 37, 77
abundances, 133
adiabatic separation, 49
aerosol, 137–139
aerosol growth, 140
aerosol science, 137
Ag clusters, 152
Ag nanoparticles, 122
aggregation, *see* gas aggregation
Aitken mode, 138
aliphatic chains, 135
alkaline, 7, 9–11, 14, 16, 35, 58, 65
amorphous carbon structures, 133
anisotropy, 92, 93
aromatic rings, 135
atmospheric particles, 137
atomic core, 116
atomic shells, 37, 79
atomic transitions, 42
atoms, 7

B
binding energy, 70
biocompatible polymers, 146
biological clearance, 156
biosensors, 149
blood-pool agents, 147
BO, *see* Born–Oppenheimer
BO molecular dynamics, 50
BO surface, 50, 51
BO-MD, *see* BO molecular dynamics

Born–Oppenheimer, 48
bulk equilibrium, 130

C
cancer therapies, 153
carbon nanotube functionalization, 149
carbon nanotubes, 149, 156
Car–Parrinello molecular dynamics, 50
catalysis, 121
catalysis and geometry, 122
catalytic efficiency, 122
CCN, *see* cloud condensation nuclei
charge resonance-enhanced ionization, 96
chemical binding, 9
chemotherapy, 151
chirp, 94, 100
chromophore, 147, 150, 152, 153
CI, *see* configuration interaction
cisplatin, 151, 155
classical, 161
Clemenger–Nilsson model, 58, 80
cloud condensation nuclei, 138, 140
cloud formation, 137
cluster deformation, 40
cluster shape, 80, 81, 85
cluster source, 25
cluster tailoring, 145
cluster temperature, 29
cluster-substrate coupling, 117
coating molecules, 105
coherent line broadening, 87
cohesive energy, 14
collective splitting, 85
collision, 32, 162
conduction gap, 120
conductivity, 11, 34
conductor, 11
configuration interaction, 56
contrast agents, 146

cooling, 51
core electron, 5, 6, 8, 35, 42, 65, 95
Coulomb explosion, 97
Coulomb Hartree energy, 60
Coulomb interaction, 83, 116
Coulomb pressure, 98
coupled oscillators, 109
covalent binding, 15
CP-MD, *see* Car–Parrinello molecular dynamics
CREI, *see* charge resonance-enhanced ionization
critical saturation ratio, 142, 143
cytotoxicity, 156

D

damage, 153
deformation, 68, 70, 86
degrees of freedom, 115
density functional theory, 8, 59
density of states, 89, 124
deposited clusters, 2, 19, 26
designer material, 125
DFT, *see* density functional theory
dimer molecule, 9
dipole antenna, 124
dipole moment, 70
dipole polarizability, 71
dipole strength, 73
dipole–dipole interaction, 109
direct emission, 74
disintegration, 105
dissociation energies, 78
distance-dependent tight binding, 53
distorted-wave Born approximation, 32
DNA, 146
DNA double strand breaks, 155
DNA sensors, 149
drug delivery, 150, 152
DWBA, *see* distorted-wave Born approximation

E

effective potential, 53
electrical conductivity, 71
electron gas, 129
electron transmission, 119
electronic shells, 37, 80
electron–electron correlations, 87
embedded clusters, 1, 2, 18, 19, 25, 26
environment, 26
exchange-correlation energy, 60
exponential decrease, 90

F

FEL, *see* free-electron lasers
Fermi momentum, 34, 62, 128
ferromagnetic clusters, 154
field-dominated regime, 90, 95
finite fermion systems, 129
fluorescence, 124
fluorescence enhancement, 124
force fields, 113
fragmented resonance, 83
free-electron lasers, 94
frequency-dominated regime, 95
fullerenes, 25, 37, 156

G

galaxy, 134
gas aggregation, 25
generalized-gradient approximation, 63
GGA, *see* generalized-gradient approximation
glass matrix, 104
global shape parameters, 70
global warming, 137
glucose sensor, 149
GNP, *see* gold nanoparticles
gold cluster functionalization, 145
gold clusters, 153, 155, 156
gold nanoparticles, 147, 150
gold nanoshell, 153, 157
gold-silica nanoparticles, 154
ground-state configuration, 51

H

Hagena parameter, 24
halogen, 7, 9, 11, 16
harmonic oscillator, 58
Hartree–Fock, 56
He droplets, 107
HF, *see* Hartree–Fock
highest occupied molecular orbitals, 8
highly oriented pyrolytic graphite, 2
HOMO, *see* highest occupied molecular orbitals
homogeneous electron gas, 63
homogeneous matter, 129
HOMO–LUMO gap, 81
HOPG, *see* highly oriented pyrolytic graphite
Hubbard model, 53
Hückel, 52
hydrodynamical flow, 98, 99
hygroscopicity, 143
hyperthermia, 154

I

independent particle picture, 60

infrared measurements, 135
inner ionization, 96
insulator, 11
interaction potentials, 116
interface, 113
interstellar objects, 135
ionic binding, 9–11, 15
ionic core, 7, 8, 47
ionic shells, 37
ionization, 43, 88, 108
ionization potential, 13, 70, 95
ionization threshold, 92
IP, see ionization potential
isomer, 12
isotropy, 93

J
Jahn–Teller effect, 80
jellium, 62, 68, 86

K
Keldysh parameter, 95
Kelvin effect, 141
Kelvin equation, 141
Köhler curve, 142
Köhler theory, 141
Kohn–Sham equations, 60
Kohn–Sham potential, 68, 78
Kohn–Sham scheme, 59
KS, see Kohn–Sham equations

L
Landau fragmentation, 83, 86
laser, 32
laser field, 32
laser frequency, 94
laser intensity, 33, 94
laser profile, 33
LCAO, see linear combination of atomic orbitals
LDA, see local density approximation
Lennard-Jones, 53
line broadening, 86
linear combination of atomic orbitals, 52
liquid ^{3}He, 129
local density approximation, 60, 62
local spin-density approximation, 62
lowest unoccupied molecular orbital, 8
LSDA, see local spin-density approximation
luminescence, 109
LUMO, see lowest unoccupied molecular orbital

M
magic numbers, 78, 80, 81, 131
magnetic moment, 38, 106
magnetic resonance, 146
magnetic resonance imaging, 146
magnetism, 106
many-body problem, 47
marker, 146
mass spectrometer, 21
mass spectrometry, 27
MC, see Monte Carlo
MCHF, see multiconfigurational HF
MD, see molecular dynamics
metabolic clearance, 144
metallic binding, 15
MgO surface, 103
Mie, 2
Mie frequency, see Mie plasmon frequency
Mie plasmon frequency, 34
Mie plasmon resonance, 81, 82, 84, 132
Mie surface plasmon, 40, 110, 117
mode coupling, 109
molecular dynamics, 53
molecular vibrations, 95
moments of inertia, 71
monomer separation energy, 70
Monte Carlo, 52
MPI, see multiphoton ionization
MRI, see magnetic resonance imaging
multiconfigurational HF, 57
multiphoton, 162
multiphoton ionization, 43, 88, 90, 95, 96
multipole moments, 70

N
nanomaterial, 125
nanomechanics, 120
nanometer droplets, 143
nanoplasma, 98
nanoplasma model, 99
nanotoxicity, 156
near-field enhancement, 124
nuclear giant resonances, 132
nuclear matter, 129
nucleation mechanism, 139
nucleation mode, 138, 139

O
oblate, 40, 80, 85
one-particle-one-hole, 56, 82
one-photon process, 43, 89, 91
ONIOM, 113
optical absorption, 72, 82

optical field ionization, 95
optical response, 40, 72, 86, 104
orientation averaging, 92
outer ionization, 96

P

PAD, *see* photoelectron angular distribution
PAH, *see* polycyclic aromatic hydrocarbon
Pauli repulsion, 116
PCR, *see* polymerase chain reaction
PES, *see* photoelectron spectroscopy
PES/PAD, 44, 92
photoabsorption, 40
photocatalysis, 122
photoelectron angular distribution, 44, 91, 119
photoelectron spectroscopy, 13, 43, 89
photon-induced ionization, 96
photosensitizers, 147
photothermal therapy, 150, 153
planetary nebulae, 135
plasmonic resonance energy transfer, 147
polarization interaction, 116
polarization potentials, 67
polycyclic aromatic hydrocarbon, 124, 134
polymer hydrogels, 151
polymerase chain reaction, 149
ponderomotive potential, 94
potential energy surface, 50
power spectrum, 73
PRET, *see* plasmonic resonance energy transfer
prodrug, 151
projectile, 32
prolate, 40, 80, 85
pseudo wave functions, 67
pseudojellium, 69
pseudomass, 50
pseudopotentials, 65
PsP, *see* pseudopotentials
pulse delay, 107
pulse length, 100
pump and probe, 107

Q

QM/MM, *see* quantum mechanical/molecular mechanical
quadrupole moment, 70, 80
quantum dots, 132
quantum mechanical/molecular mechanical, 112

R

radiative forcing, 137
radiosensitizer, 150
radiotherapy, 155
random-phase approximation, 63, 72
Raoult's law, 141
rare gas clusters, 100
rare gas matrix, 103, 117
recoil energies, 98
reflectron, 38
relative humidity, 140
relaxation, 5, 6, 52, 74
residual interaction, 83
resonance conditions, 100
resonance mechanism, 100
resonance width, 82, 83
Rh nanostructures, 106
RNA, 146
root-mean-square radius, 70
RPA, *see* random-phase approximation

S

saturating fermion systems, 127
saturation density, 128
saturation ratio, 140
saturation vapor pressure, 141
self-interaction correction, 63, 70
semiclassical approaches, 64, 79
sensor, 147
shell closures, 77, 131
shell oscillations, 131
shell structure, 7
short-range potential, 117
short-range repulsion, 116
SIC, *see* self-interaction correction
single-electron energy, 36, 43, 67, 70, 80
Slater approximation, 63
Slater determinant, 56
Slater exchange energy functional, 63
soft jellium, 68
solute molecule, 141
spectral analysis, 73
spectral fragmentation, 73
SPION, *see* superparamagnetic ion oxide nanoparticles
splitting, 85
superatoms, 125
supernova, 133
supernova ejecta model, 133
superparamagnetic ion oxide nanoparticles, 146
supersaturation, 141
supershells, 79
supersonic jet, 21, 23
surface plasmon, *see* Mie surface plasmon
surface plasmon resonance, 147

surface source, 25
surface states, 122

T
TDDFT-MD, 51
TDHF, *see* time-dependent HF
TDLDA, *see* time-dependent local density approximation
TDLDA-MD, 87
therapeutic window, 151
thermal broadening, 83, 87
thermal emission, 74, 89
thermal shape fluctuations, 86
Thomas–Fermi approximation, 64
tight binding, 52
time of flight, 21, 37
time scales, 6
time-dependent HF, 56
time-dependent KS equations, 60
time-dependent local density approximation, 63
time-resolved analysis, 107
TOF, *see* time of flight
toxicity, 156
trajectory surface hopping, 51
transferability, 66
triaxial, 81, 85, 86

U
ultrafine aerosol network, 138
ultrafine aerosols, 138
unidentified infrared bands, 134, 135

V
valence electron, 5, 8, 40, 42, 43, 47, 65, 97, 116
van der Waals binding, 10, 11, 15
van der Waals cluster, 15, 53
vapor phase, 141
vascular targeting, 145
velocity map imaging, 44, 92
visible light, 95
Vlasov equation, 64
Vlasov–Uehling–Uhlenbeck, 64
VMI, *see* velocity map imaging
VUU, *see* Vlasov–Uehling–Uhlenbeck

W
water activation, 138, 140, 143
water activity, 141
water liquid droplet, 141
water molecules, 105
Weisskopf formula, 74
Wigner–Seitz, 128
Wigner–Seitz radius, 35, 62
Woods–Saxon shell model, 58

X
XMCD, *see* X-ray magnetic circular dichroism
X-ray magnetic circular dichroism, 107
X-ray regime, 95
X-ray spectrum, 42

Z
ZEKE, *see* zero electron kinetic energy
zero electron kinetic energy, 43